HANDBOOK OF ELECTRONIC MATERIALS
Volume 7

HANDBOOK OF ELECTRONIC MATERIALS

Compiled by:

ELECTRONIC PROPERTIES INFORMATION CENTER
Hughes Aircraft Company
Culver City, California

Sponsored by:

U.S. DEFENSE SUPPLY AGENCY
Defense Electronics Supply Center
Dayton, Ohio

Volume 1:
OPTICAL MATERIALS PROPERTIES, 1971

Volume 2:
III-V SEMICONDUCTING COMPOUNDS, 1971

Volume 3:
SILICON NITRIDE FOR MICROELECTRONIC APPLICATIONS, PART I:
                                        PREPARATION AND PROPERTIES, 1971

Volume 4:
NIOBIUM ALLOYS AND COMPOUNDS, 1972

Volume 5:
GROUP IV SEMICONDUCTING COMPOUNDS, 1971

Volume 6:
SILICON NITRIDE FOR MICROELECTRONIC APPLICATIONS, PART II:
                                        APPLICATIONS AND DEVICES, 1972

Volume 7:
III-V TERNARY SEMICONDUCTING COMPOUNDS—DATA TABLES, 1972

HANDBOOK OF ELECTRONIC MATERIALS
Volume 7

III-V Ternary Semiconducting Compounds–Data Tables

M. Neuberger

Electronic Properties Information Center
Hughes Aircraft Company, Culver City, California

IFI/PLENUM · NEW YORK-WASHINGTON-LONDON · 1972

This document has been approved for public release and sale;
its distribution is unlimited. Sponsored by U.S. Defense Supply
Agency, Defense Electronics Supply Center, Dayton, Ohio.
Under Contract No. DSA 900-72-C-1182

Library of Congress Catalog Card Number 76-147312
ISBN 978-1-4684-6167-1   ISBN 978-1-4684-6165-7 (eBook)
DOI 10. 1007/978-1-4684-6165-7

CONTENTS

	Page
INTRODUCTION	1
GALLIUM-ALUMINUM-ANTIMONY SYSTEM	5
Bibliography	7
GALLIUM-ALUMINUM-ARSENIC SYSTEM	8
Bibliography	11
GALLIUM-ALUMINUM-PHOSPHOROUS SYSTEM	13
Bibliography	14
GALLIUM-ARSENIC-ANTIMONY SYSTEM	15
Bibliography	18
GALLIUM-ARSENIC-PHOSPHOROUS SYSTEM	19
Bibliography	25
GALLIUM-INDIUM-ANTIMONY SYSTEM	28
Bibliography	33
GALLIUM-INDIUM-ARSENIC SYSTEM	35
Bibliography	39
GALLIUM-INDIUM-PHOSPHOROUS SYSTEM	41
Bibliography	44
INDIUM-ARSENIC-ANTIMONY SYSTEM	46
Bibliography	49
INDIUM-ARSENIC-PHOSPHOROUS SYSTEM	51
Bibliography	55

INTRODUCTION

The Electronic Properties Information Center has developed the Data Table as a compilation of the most reliable information available for the physical, crystallographic, mechanical, thermal, electronic, magnetic and optical properties of a given material. Data Tables formerly served as an introduction to the graphic data compilations on the material published by the Electronic Properties Information Center, EPIC, as Data Sheets. Although the Data Sheets were principally concerned, according to the scope of the Center, with electronic and optical data, it is believed that data covering the complete property spectrum, is of the first importance to every scientist and engineer, whatever his information requirements. The enthusiastic reception of these Data Tables has confirmed this opinion and increasing requests for this highly selective type of information resulted in the publication of volume 2 in this series, "III-V Semiconducting Compounds" in 1971 and "Group IV Semiconducting Compounds", also in 1971. Recent interest in the device applications of the ternary semiconductors, led to the compilation of these Data Tables on the III-V Ternary Semiconducting Compounds.

The major problem in this type of selective data compilation on a semiconducting material, lies in the material specifications. Properties may vary so widely with doping, crystallinity, defects, geometric forms and the other parameters of preparation, that any attempts at comparison normally fail. On this basis, we have consistently attempted to give the preparation methods, carrier concentrations, and physical form. At the very least, these data should be reproducible and this gives the data their principal validity. If such values however, are not available, then the next best data are reported, together with material specifications.

Values for a range of temperatures, wavelengths, frequencies, pressures and field strengths (both electric and magnetic), are reported where available. Our primary goal has been not to compress, but to select and present a rounded and fully representative view of the specific material.

This comprehensive review of each compound has been made possible by the extensive collection of documents in the EPIC files; to date 50,000 technical journal articles and government reports have been acquired by the Center. To compile these III-V Ternary Semiconducting Compounds Data Tables, about 350 of these documents, reported on one or more of the III-V Ternary Compounds, have been evaluated for relevant data.

As far as possible, the arrangement of data has been standardized in a consistent order as follows:

PHYSICAL, MECHANICAL, THERMAL

Property	Unit
Formula	
Molecular Weight	
Density	g/cm^3
Name	
Mineral Name	
Color	
Hardness	Mohs, kg/mm^2
Cleavage	
Symmetry	
Space Group	
Lattice Parameters	$\overset{\circ}{A}$
Melting Point	$^\circ C$
Sublimation Temperature	
Specific Heat	$cal/g^\circ K$
Debye Temperature	$^\circ K$
Thermal Conductivity	$W/cm^\circ K$
Thermal Expansion Coefficient	$10^{-6}/^\circ K$
Elastic Coefficient	
Compliance, s	$cm^2/dyne$
Stiffness or Elastic Modulus, c	$dyne/cm^2$

Property	Unit
Shear Strength	kg/cm^2
Young's Modulus	$dynes/cm^2$
Poisson's Ratio	
Sound Velocity	cm/sec.
Compressibility (1/Bulk Modulus)	$cm^2/dyne$

ELECTRICAL, ELECTRONIC

Dielectric Constant	
Static, ε_0	
Optic ε_∞	
Dissipation Factor, tg δ	
Electrical Resistivity	ohm-cm
Mobility	cm^2/V sec
Electron, μ_n	
Hole, μ_p	
Temperature Coefficient, T^x	
Lifetime, τ	sec.
Piezoelectric Coefficients	C/N, C/m^2, m/V
Piezoresistance Coefficients	$cm^2/dyne$
Elastoresistance Coefficients	$cm^2/dyne$
Effective Mass	
Energy Levels	eV
Temperature Coefficient, dE/dT	$eV/°K$
Pressure Coefficient, dE/dP	$eV/kg\ cm^{-2}$
Field Coefficient	
Dilatation Coefficient	
Deformation Potential	eV
Photoelectric Threshold, Φ	eV
Work Function, ϕ	eV
Electron Affinity, ψ	eV
Barrier Heights	eV
Phonon Spectra	meV
Seebeck Coefficient	$V/°K$
Nernst-Ettingshausen Coefficient	
Magnetic Susceptibility	10^{-7} cgs
g-Factor	

OPTICAL

Transmission	%
Refractive Index	
Temperature Coefficient	$°K^{-1}$
Spectral Emissivity	
Piezo-optic Coefficient	$cm^2/dyne$
Elasto-optic Coefficient	
Electro-optic Coefficient	
Laser Properties	Å

Changes in value with temperature and pressure are always given where available. The units have been standardized as far as possible in the cgs system, except for piezoelectric coefficients which, according to the usage in this country, are given in Coulombs/Newton; certain mechanical properties data are given in psi.

The most highly valued aspect of this work is the fact that every individual data point is accompanied by a reference citation. Many data compilations appear in the literature which contain little or no documentation as to the data sources; this work allows the reader to refer to the original research paper for additional information and in this way offers a representative bibliographic review of the III-V compounds. Where two or more documents present the same data values, all are cited. The bibliography which follows every set of tables is arranged alphabetically by author; more than one document by the same author is distinguished by the letters A, B, C, etc. In order to keep each compound separately, many pages have been left with a considerable amount of blank space. This should prove useful however, in furnishing the reader with space for the addition of the latest information as it appears in the published literature. In a few cases, citations are out of order because they were added in proof.

The importance of several of these ternary compounds in device applications, has suggested that various device properties should be added to the compilation. Laser properties, Light-Emitting Diode properties and electron emission devices (cold cathodes) are included.

Grateful acknowledgement is made to the careful review of these tables by Dr. Victor Rehn of the U.S. Naval Ordnance Test Station, Michelson Laboratory at China Lake, California and Dr. Wolfgang Klein of the U.S. Army Electronics Command, Night Vision Laboratory at Fort Belvoir, Virginia. All errors and omissions however, are solely the responsibility of the author.

GALLIUM-ALUMINUM-ANTIMONY SYSTEM

PROPERTY	SYMBOL	VALUE		UNIT	NOTES	TEMP.(°K)	REFERENCES
Formula		$Al_xGa_{1-x}Sb$					
Symmetry		cubic					
Lattice Parameter	a_o	x	a_o				
		0	6.094	Å	GaSb		Donnay
		4	6.1007		single phase, zone cast		Miller et al.
		55	6.1201		ingot		
		62	6.1200				
		74	6.1237				
		100	6.1355		InSb		Giesecke & Pfister
Knoop Hardness	H_{100}	50	335	kg/mm^2	polycrystalline		Miller et al.
	H_{50}	10-100	420-450		polycrystalline, homogeneity increases with zone refining		Burdiyan
Melting Point	M.P.		M.P.				
		0	712.1	°C	GaSb		Bednar & Smirous
		25	860				Miller et al.
		50	955				
		75	1010				
		100	1080		AlSb		Glazov et al.
Energy Gap	E_g		E_g				
		0	0.70		GaSb optical meas.	300	Cardona
		4	0.74		optical meas.	300	Miller et al.
		55	1.38		single phase, zone cast		
		62	1.43		ingot $n_p= 1-4\times10^{17}$ cm^{-3}		
		74	1.46				
		20	0.94		optical meas. on cast	300	Burdiyan & Kolomiets
		40	1.16		ingot, $n_p= 8.6\times10^{17}$ to		
		50	1.24		2.8×10^{18} cm^{-3}		
		80	1.49				
		100	2.218 (direct)		AlSb optical meas.	300	Cardona
			1.62 (indirect)		optical meas.	300	Oswald & Schade

Energy Band Structure	x	E_o	E_1	$E_1+\Delta_1$	E_2			
	0	0.7	2.05	2.7	4.2	eV	GaSb	300	Kroitoru et al.
	20		2.1	2.8	4.2		reflectivity meas. on		
	40		2.5	2.85	4.2		polycrystalline alloy		
	50		2.5	2.85	4.25				
	60		2.6	2.9	4.3				
	100	1.55	2.85	3.2	4.45		AlSb		

5

GALLIUM-ALUMINUM-ANTIMONY SYSTEM

PROPERTY	SYMBOL	VALUE		UNIT	NOTES	TEMP.(°K)	REFERENCES
Mobility	μ	x	μ				
		4	357	cm^2/V sec	$n_p = 1\text{-}10\times10^{17}$ cm^{-3}	300	Miller et al.
		55	264				
		62	264				
		74	264				
		20	75		$n_p = 8.6\text{-}28\times10^{17}$	300	Burdiyan & Kolomiets
		80	525				
Electrical Resistivity	ρ		ρ				
		4	0.014	ohm-cm		300	Miller et al.
		55	0.075				
		62	0.11				
		74	0.22				

x	110°K	330°K	900°K				
20	10^{-4}	2×10^{-4}	1.7×10^{-3}	Ω-cm	single phase, polycrystalline		Burdiyan & Kolomiets
50	2×10^{-3}	10^{-3}	10^{-2}				
80	7×10^{-4}	4×10^{-4}	3.0×10^{-3}				

BEDNAR, J. and K. SMIROUS. The Melting Point of Gallium and Indium Antimonide (In Ger.). CZECH. J. OF PHYS., v. 5, no. 4, 1955. p. 546.

BURDIYAN, I.I. Additional Evidence Concerning Solid Solutions in the Aluminum Antimonide-Gallium Antimonide System. SOVIET PHYS. SOLID STATE, v. 1, no. 9, Mar. 1960. p. 1246-1252.

BURDIYAN, I.I. and B.G. KOLOMIETS. Investigation of the Conductivity and the Hall Effect in the Aluminum Antimonide-Gallium Antimonide System. In: ALL-UNION CONFERENCE ON SEMICONDUCTOR MATERIALS, PROC. OF THE 4TH. Edited by N.Kh. Abrikosov. New York, Consultants Bureau, 1961. p. 101-102.

CARDONA, M. Fundamental Reflectivity Spectrum of Semiconductors with Zincblende Structure. J. OF APPLIED PHYS., Supplement to v. 32, no. 10, Oct. 1961. p. 2151-2155.

DONNAY, J.D.H. (Editor). Crystal Data. Determinative Tables. 2nd Edition. American Crystallographic Association. Apr. 1963. ACA Monograph no. 5.

GIESECKE, G. and H. PFISTER. Precision Determination of the Lattice Constants of III-V Compounds (In Ger.). ACTA CRYSTALLOGRAPHICA, v. 11, 1958. p. 369-371.

GLAZOV, V.M. et al. Liquid Semiconductors. New York, Plenum Press, 1969. 362 p.

KROITORU, S.G. et al. Energy Structure of Several Solid State Materials on a Basis of Combinations of Groups III-V (In Russ.). AKAD. NAUK SSSR IZV. NEORGAN. MAT., v. 2, no. 5, 1966. p. 805-809.

MILLER, J.F. et al. Preparation and Properties of Aluminum Antimonide-Gallium Antimonide Solid Solution Alloys. ELECTROCHEM. SOC., J., v. 107, no. 6, June 1960. p. 527-533.

OSWALD, R. and R. SCHADE. On the Determination of the Optical Constants of III-V Semiconductors in the Infrared (In Ger.). Z. FUER NATURFORSCH., v. 9a, no. 7/8, July/Aug. 1954. p. 611-617.

PROPERTY	SYMBOL	VALUE		UNIT	NOTES	TEMP.(°K)	REFERENCES
Formula		$Ga_xAl_{1-x}As$					
Density		x	g/cm^3				
		0	3.598		AlAs	300	Donnay
		34	4.29		closed tube, iodine		Black & Ku
		42	4.40		vapor transport, single		
		95	5.24		crystals deposited on		
					high purity, (110) GaAs		
		100	5.307		GaAs		Bateman et al.
Color		10	orange		transmitted white light		Manasevit,
		20	red-orange		through epitaxial, CVD		Bindeman et al.
		60	red		thin films on alumina		
		70	reddish black		$\sim 4\mu$ thick		
		80	black				
Symmetry		cubic					
Lattice Parameter	a_o		a_o (Å)				
		0	5.6605		AlAs		Ettenberg & Paff
		42	5.6581				Black & Ku
		100	5.65191		GaAs		Cooper
Thermal Expansion Coeff.		0	5.20	$10^{-6}/°C$	AlAs, linear from	20°C	Ettenberg & Paff
					20-1000°C, lattice		
					match AlAs-GaAs at		
					800-1000°C complete		
		100	6.86		GaAs		Pierron et al.

Liquidus Isotherms	At. % of Liquidus						
	Ga	Al	As	T°C			
	96.5	1.0	2.5	898	first solid for slow		Panish &
	94.0	1.0	5.0	952	cooling of high gallium		Sumski,
	89.0	1.0	10.0	1037	solutions		Ilegems &
	84.0	1.0	15.0	1082			Pearson (B)
	92.5	5.0	2.5	1002			
	90.0	5.0	5.0	1067			
	85.0	5.0	10.0	1140			
	82.5	15.0	2.5	1074			

PROPERTY	SYMBOL	VALUE		UNIT	NOTES	TEMP.(°K)	REFERENCES
Dielectric Constant Optical	ε_∞	x	ε_∞				
		90-99.6	11.0		optical meas.	300	Sikharulidze
					n-type, polycrystalline		et al.
		18	8.5		reflectivity meas. on	300	Ilegems & Pearson (A)

Effective Mass Electron	m_n	x	m_n	$n(10^{18}\ cm^{-3})$			
		90	0.071	1.32	reflectivity meas. on	300	Sikharulidze
		94	0.074	1.60	n-type, polycrystalline		et al.
		96.5	0.070	1.63	material, 30μ thick		
		99.6	0.064	1.80			

GALLIUM-ALUMINUM-ARSENIC SYSTEM

PROPERTY	SYMBOL	VALUE		UNIT	NOTES	TEMP.(°K)	REFERENCES

Energy Gap — E_g (Direct E_{gd}, Indirect E_{gi})

x	$E_{gd}(\Gamma_{15}-\Gamma_1)$	$E_{gi}(\Gamma_{15}-X_1)$	UNIT	NOTES	TEMP.(°K)	REFERENCES
0	2.90	2.13	eV	AlAs	300	Kischio, Lorenz et al.
25	2.42	2.0		Schottky barrier photoresponse meas.	300	Casey & Panish
34	2.30	1.96				
48	2.12	1.86				
55	2.0	1.85				
68	1.82	-				
85	1.6	-				
100	1.4257			GaAs	300	Zvara

Energy Band Structure

x	E_{gd}	E_{gi}	E_1	$E_1+\Delta_1$	UNIT	NOTES	TEMP.(°K)	REFERENCES
53		1.94	3.15	3.33	eV	molecular beam, vapor preparation, single crystal films, reflectivity meas.	300	Cho & Stokowsk
57	1.97		3.06	3.27				
75	1.81							
80	-		2.97	3.16				
83	1.67		2.96	3.14				
87.5	-		2.93	3.14				
90	-		2.92	3.13				

x	E_o	$E_o+\Delta_o$	E_1	$E_1+\Delta_1$	$E_o{}'$	$E_o{}'+\Delta_o{}'$	E_2	NOTES	TEMP.(°K)	REFERENCES
0	2.93	2.95						AlAs	300	Berolo & Woolley
25	2.49	2.54	3.5	3.7	4.7	4.75	4.85	electroreflectance meas. on		
34	2.36	2.39	3.4	3.6	4.7	4.75	4.85	LPE deposited,		
48	2.16	2.19	3.25	3.5	4.7	4.7	4.9	1 mil thick layers		
55	2.04	2.06	3.2	3.4	-		4.9	on GaAs		
68	1.80	1.83	-	-	-		4.9			
85	1.63	1.66	2.91	3.0	4.4	4.6	5.0			
100	1.42	1.45	2.9	3.0	4.4	4.6	5.0	GaAs		

Direct-Indirect Cross-over

x	VALUE	NOTES	TEMP.(°K)	REFERENCES
64	1.92 eV	electroluminescence meas.	300	Dierschke et al.
65	1.92	electroluminescence meas.	300	Berolo & Woolley

Phonon Branch Spectra — Longitudinal Optic LO, Transverse Optic TO

x	LO_1	LO_2	TO_1	TO_2	UNIT	NOTES	TEMP.(°K)	REFERENCES
0	49.60	-	44.89	-	meV	reflectivity meas. on single crystals	300	Ilegems & Pearson (A)
18	49.60	45.25	44.89	31.74				
47	47.87	33.48	44.89	31.98				
55	47.61	34.46	44.63	32.24				
59	47.10	35.46	43.65	32.37				
62	46.75	35.71	44.15	32.49				
68	46.36	35.96	44.15	32.99				
79	44.89	36.02	43.65	32.37				
92	44.63	36.08	44.15	33.11				
100	-	36.21	-	32.24				

Refractive Index — n

x	n	NOTES	TEMP.(°K)	REFERENCES
18	2.9	reflectivity meas. on single crystals	300	Ilegems & Pearson (A)
90	3.3	optical meas. on single crystals	300	Sikharulidze et al.

GALLIUM-ALUMINUM-ARSENIC SYSTEM

PROPERTY	SYMBOL	VALUE					UNIT	NOTES	TEMP.(°K)	REFERENCES

Electron Emission (Cold Cathode)

	x	Wavelength (Å)	Emission Current Density (A/cm^2)	Efficiency (%)	Photo-sensitivity (µA/lm)	NOTES	TEMP.(°K)	REFERENCES
	88	8700-8800	0.1 / 7.0	4.0 / 1.1	700-1000	LPE-deposited on doped GaAs, Cs$_2$O-covered	300	Schade et al. (A, B)

Use in a D.H. Junction Laser

x	Wavelength (Å)	TCD A/cm^2	Efficiency (%)	Power Output	NOTES	TEMP.(°K)	REFERENCES
75	8700	5x10^3	46-47	200 mW	double heterostructure junction laser, LPE, substrate: GaAs, n-type, Si-doped 1)n-, Sn-doped, GaAlAs, 10^{17} 2)n-,p-, doped GaAs, 1-8x10^{18} 3)p-, Ge-doped, GaAlAs, 5x10^{17} 4)p-, Ge-doped, GaAs, 5x10^{18} layers 0.5-1µ thick	300	Pinkas et al., Miller et al. (A, B)
	9460 / 9130	24 / 1100	50		large optical cavity, heterojunction laser diode, 400µ long, 2µ thick	300	Kressel et al.
		3600 / 2000		0.04 W / 1.2 W	continuous wave oper. pulsed operation	300	
60-80 / 94	8576 / 8540	1000 / 2500	30-40	20 mW	double heterostructure LPE injection laser, continuous operation; emits polarized light	311	Hayashi et al.

Electroluminescent Diodes — Quantum Efficiency η

x	Wavelength (Å)	Current Density	η (%)	Power Output	NOTES	TEMP.(°K)	REFERENCES
	8150 / 6950	300 mA	14 / 4	60 mW	Zn-diffused diodes, LPE deposited on GaAs 10-15µ junction depth	300	Dierschke et al.
	6700		0.3 / 4.0		LPE deposited on GaAs n-, p-type layers, 2-5µ thick	300 / 77	Beneking et al.
	9300				Al$^+$ ion implantation of Zn-doped GaAs, 0.2µ thick	77	Hunsperger & Marsh
	6700-7000				annealed 5 hr. at 900°C	77	

x	Wavelength (Å)	TCD A/cm^2	Efficiency (%)	Luminance	NOTES	TEMP.(°K)	REFERENCES
62-67	6650	40	0.23	10^4 ft L	LPE deposited p-n junction, 5-7µ thick	300	Shih & Blum

x	Wavelength (Å)			Power	NOTES	TEMP.(°K)	REFERENCES
70	7750-7940	7500		1.7 mW	double heterostructure, 1µ thick, diode coupled to multimode optical fibers, 2000 hour operating life	300	Burrus & Miller
10	5760 (strong)				Bi-doped single crystals, photo-luminescence meas.	4.2	Bindemann et al.

Light Modulation

x	Wavelength	Phase Modulation	Bias Voltage		NOTES	TEMP.(°K)	REFERENCES
70	11530	180°	10 V	0.1 mW/1 MHz	1 mm length diode	300	Reinhart & Miller

BATEMAN, T.B. et al. Elastic Moduli of Single Crystal Gallium Arsenide. J. OF APPLIED PHYS., v. 30, no. 4, Apr. 1959. p. 544-545.

BEROLO, O. and J.C. WOOLLEY. Electroreflectance Spectra of Aluminum Gallium Arsenide Alloys. CANADIAN J. OF PHYS., v. 49, no. 10, May 1971. p. 1335-1339.

BENEKING, H. et al. Improved Technique for the Preparation of $Ga_xAl_{1-x}As$ Electroluminescent Diodes. ELECTRONICS LETTERS, v. 8, no. 1, Jan. 13, 1972. p. 16-17.

BINDEMANN, R. et al. Photoluminescence of Bi-Doped $Al_{1-x}Ga_xAs$ Single Crystals. PHYS. STATUS SOLIDI A, v. 7, no. 2, Oct. 16, 1971. p. K121-K123.

BLACK, J.F. and S.M. KU. Preparation and Properties of Aluminum Arsenide-Gallium Arsenide Mixed Crystals. ELECTROCHEM. SOC., J., v. 113, no. 3, Mar. 1966. p. 249-254.

BURRUS, C.A. and B.I. MILLER. Small-Area, Double-Heterostructure Aluminum-Gallium Arsenide Electroluminescent Diode Sources for Optical-Fiber Transmission Lines. OPTICS COMMUNICATIONS, v. 4, no. 4, Dec. 1971. p. 307-309.

CASEY, H.C. and M.B. PANISH. Composition Dependence of the Gallium Aluminum Arsenide Direct and Indirect Energy Gaps. J. OF APPLIED PHYS., v. 40, no. 12, Nov. 1969. p. 4910-4912.

CHO, A.Y. and S.E. STOKOWSK. Molecular Beam Epitaxy and Optical Evaluation of Aluminum Gallium Arsenide. SOLID STATE COMMUNICATIONS, v. 9, no. 9, May 1971. p. 565-568.

COOPER, A.S. Precise Lattice Constants of Germanium, Aluminum, Gallium Arsenide, Uranium, Sulfur, Quartz and Sapphire. ACTA CRYSTALLOGRAPHICA, v. 15, 1962. p. 578-582.

DIERSCHKE, E.G. et al. Efficient Electroluminescence from Zinc-Diffused $Ga_{1-x}Al_xAs$ Diodes at 25°C. APPLIED PHYS. LETTERS, v. 19, no. 4, Aug. 15, 1971. p. 98-100.

DONNAY, J.D.H. (Ed.) Crystal Data. Determinative Tables. 2nd Ed. American Crystallographic Association, Apr. 1963. ACA Monograph no. 5.

ETTENBERG, M. and R.J. PAFF. Thermal Expansion of AlAs. J. OF APPLIED PHYS., v. 41, no. 10, Sept. 1970. p. 3926-3927.

HAYASHI, I. et al. $GaAs-Al_xGa_{1-x}As$ Double Heterostructure Injection Lasers. J. OF APPLIED PHYS., v. 42, no. 5, Apr. 1971. p. 1929-1941.

HUNSPERGER, R.G. and O.J. MARSH. $Ga_{1-x}Al_xAs$ Produced by Al^+ Ion Implantation of GaAs. APPLIED PHYS. LETTERS, v. 19, no. 9, Nov. 1, 1971. p. 327-329.

ILEGEMS, M. and G.L. PEARSON. Infrared Reflection Spectra of $Ga_{1-x}Al_xAs$ Mixed Crystals. PHYS. REV., B, Ser. 3, v. 1, no. 4, Feb. 15, 1970. p. 1576-1582. [A]

ILEGEMS, M. and G.L. PEARSON. Derivation of the Ga-Al-As Ternary Phase Diagram with Applications to Liquid Phase Epitaxy. PROC. SECOND INT. SYMP. ON GaAs (IPPS, London, 1969), pp. 3-10. [B]

KISCHIO, K. Aluminum Arsenide (In Ger.). Z. FUER ANORG. UND ALLGEM. CHEM., v. 328, 1964. p. 187-193.

KRESSEL, H. et al. Large-Optical-Cavity (AlGa)As-GaAs Heterojunction Laser Diode: Threshold and Efficiency. J. OF APPLIED PHYS., v. 43, 1972. p. 561-567.

LORENZ, M.R. et al. The Fundamental Absorption Edge of Aluminum Arsenide and Aluminum Phosphide. SOLID STATE COMMUNICATIONS, v. 8, no. 9, May 1970. p. 693-697.

MANASEVIT, H.M. The Use of Metal-Organics in the Preparation of Semiconductor Materials. ELECTROCHEM. SOC., J., v. 118, no. 4, Apr. 1971. p. 647-650.

MILLER, B.I. et al. Reproducible Liquid-Phase-Epitaxial Growth of Double Heterostructure $GaAs-Al_xGa_{1-x}As$ Laser Diodes. J. OF APPLIED PHYS., v. 43, no. 6, June 1972. p. 2817-2826. [A]

MILLER, B.I. et al. Highly Uniform $Al_xGa_{1-x}As$ Double-Heterostructure Lasers and Their Characteristics at Room Temperature. APPLIED PHYS. LETTERS, v. 19, no. 9, Nov. 1, 1971. p. 340-343. [B]

PANISH, M.B. and S. SUMSKI. Ga-Al-As:Phase, Thermodynamic and Optical Properties. J. OF PHYS. AND CHEM. OF SOLIDS, v. 30, 1969. p. 129-137.

PIERRON, E.D. et al. Coefficient of Expansion of Gallium Arsenide, Gallium Phosphide, and Gallium Arsenic Phosphide Compounds from 62 to 200°C. J. OF APPLIED PHYS., v. 38, no. 12, Nov. 1967. p. 4669-4671.

PINKAS, E. et al. GaAs-Al$_x$Ga$_{1-x}$As Double Heterostructure Lasers-Effect of Doping on Lasing Characteristics of GaAs. J. OF APPLIED PHYS., v. 43, no. 6, June 1972. p. 2827-2835.

REINHART, F.K. and B.I. MILLER. Efficient GaAs-Al$_x$Ga$_{1-x}$As Double-Heterostructure Light Modulators. APPLIED PHYS. LETTERS, v. 20, no. 1, Jan. 1972. p. 36-38.

SCHADE, H. et al. Novel GaAs-(AlGa)As Cold-Cathode Structure and Factors Affecting Extended Operation. APPLIED PHYS. LETTERS, v. 20, no. 10, May 15, 1972. p. 385-387. [A]

SCHADE, H. et al. Efficient Electron Emission from GaAs-Al$_{1-x}$Ga$_x$As Optoelectronic Cold-Cathode Structures. APPLIED PHYS. LETTERS, v. 18, no. 10, May 15, 1971. p. 413-414. [B]

SHIH, K.K. and J.M. BLUM. Al$_x$Ga$_{1-x}$As Grown-Diffused Electroluminescent Planar Monolithic Diodes. J. OF APPLIED PHYS., v. 43, no. 7, July 1972. p. 3094-3097.

SIKHARULIDZE, G.A. et al. Optical Phenomena in Gallium Arsenide-Aluminum Arsenide Solid Solutions. SOVIET PHYS. SEMICONDUCTORS, v. 5, no. 8, Feb. 1972. p. 1302-1306.

ZVARA, M. Faraday Rotation and Faraday Ellipticity in the Exciton Absorption Region of Gallium Arsenide. PHYS. STATUS SOLIDI, v. 36, no. 2, Dec. 1969. p. 785-792.

GALLIUM-ALUMINUM-PHOSPHORUS SYSTEM

PROPERTY	SYMBOL	VALUE		UNIT	NOTES	TEMP.($^\circ$K)	REFERENCES
Formula		$Al_xGa_{1-x}P$					
Lattice Parameter	a_o	x	a_o				
		0	5.4495	Å	GaP		Pierron et al.
		100	5.4625		AlP		Richman

Melting Point	M.P.	Composition (%)					
		Ga	Al	P	M.P. ($^\circ$C)	liquid phase epitaxy, gallium-rich solution	Panish et al.
		98.3	0.7	1.0	1040		
		97.0	2.0	1.0	1102		
		94.0	5.0	1.0	1180		
		96.8	0.7	2.5	1138		
		95.5	2.0	2.5	1190		

Energy Gap	E_g	x	E_g				
		0	2.261	eV	GaP (indirect)	300	Panish & Casey
			2.33			30	Lorenz et al.
		∿25	2.28-2.31		liquid phase epitaxy diodes, electroluminescence meas.	300	Kressel & Ladany
		100	2.43		single crystal of AlP	300	Kressel & Ladany

Quantum Efficiency	η	x	η	Wavelength(μ)			
		3	7×10^{-5}	∿0.5	electroluminescence in p-n junction, Zn, 0-doped at 10 A/cm^2	2	Logan et al.
		12		∿0.6	electroluminescence in heterojunction at 0.1 A/cm^2	295	Arseni et al.

ARSENI, K.A. et al. $Al_xGa_{1-x}P$-GaP Heterojunctions. SOVIET PHYS. SEMICONDUCTORS, v. 3, no. 6, Dec. 1969. p. 788-789.

KRESSEL, H. and I. LADANY. Electroluminescence in $Al_xGa_{1-x}P$ Diodes Prepared by Liquid-Phase Epitaxy. J. OF APPLIED PHYS., v. 39, no. 11, Oct. 1968. p. 5339-5340.

LOGAN, R.A. et al. Electroluminescence in Gallium Arsenic Phosphide, Indium Gallium Phosphide, and Aluminum Gallium Phosphide Junctions. J. OF APPLIED PHYS., v. 42, no. 6, May 1971. p. 2328-2335.

LORENZ, M.R. et al. Band Gap of Gallium Phosphide from 0 to 900°K and Light Emission from Diodes at High Temperatures. PHYS. REV., v. 171, no. 3, July 1968. p. 876-881.

MERZ, J.L. and R.T. LYNCH. Preparation and Optical Properties of $Al_xGa_{1-x}P$. J. OF APPLIED PHYS., v. 39, no. 4, Mar. 1968. p. 1988-1993.

PANISH, M.B. et al. Phase and Thermodynamic Properties of the Ga-Al-P System: Solution Epitaxy of $Ga_xAl_{1-x}P$ and AlP. AIME METAL. SOC., TRANS., v. 245, no. 3, Mar. 1969. p. 559-563.

PANISH, M.B. and H.C. CASEY, JR. Temperature Dependence of the Energy Gap in Gallium Arsenide and Gallium Phosphide. J. OF APPLIED PHYS., v. 40, no. 1, Jan. 1969. p. 163-167.

PIERRON, E.D. et al. Coefficient of Expansion of Gallium Arsenide, Gallium Phosphide and Gallium Arsenic Phosphide Compounds from 62 to 200°C. J. OF APPLIED PHYS., v. 38, no. 12, Nov. 1967. p. 4669-4671.

RCA. DAVID SARNOFF RES. CENTER. Synthesis and Characterization of Electronically Active Materials. By: RICHMAN, D. TR no. 1, May 15, 1963-Feb. 15, 1964. Contract no. SD182. Mar. 15, 1964. AD 432 272.

GALLIUM-ARSENIC-ANTIMONY SYSTEM

PROPERTY	SYMBOL	VALUE		UNIT	NOTES	TEMP.(°K)	REFERENCES
Formula		$Ga_x As_{1-x} Sb$					
Symmetry		cubic					
Lattice Parameter	a_o	x	a_o	Å			Straumanis & Kim
		0	6.09592		GaSb		
		10	6.050		homogeneously melted		
		15	6.035		and annealed below		
		20	6.010		760°C		
		35	5.940				
		60	5.750				
		70	5.735				
		90	5.675				
		100	5.65321		GaAs		
		25	5.99		vacuum evaporated films		Potter & Stierwalt
		55	5.88				
		65	5.81				
Density		0	5.6137	g/cm^3	GaSb	300	McSkimin et al.
		0	5.650		hot-pressed and sintered		Sahm & Pruss
		5	5.628		powders		
		15	5.610				
		25	5.697				
		35	5.570				
		45	5.465				

PROPERTY	SYMBOL	x	M.P. Liquidus	M.P. Solidus	UNIT	NOTES	TEMP.(°K)	REFERENCES
Melting Point	M.P.	0	712.1		°C	GaSb		Bednar & Smirous
		20	990					Foster & Woods
		30	1043					
		40	1085					
		50	1123					
		60	1154					
		72	1180					
		80	1200					
		86		790				Antypas & James, Woolley
		95		900				
		97		980				
		99		1060				
		100		1238		GaAs		Richman

PROPERTY	SYMBOL	x	k(W/cm °K)	NOTES	TEMP.(°K)	REFERENCES
Thermal Conductivity	k	0	0.35	single crystal	300	Le Guillou & Albany
Lattice		0	0.120	hot pressed	300	Sahm & Pruss
		5	0.090			
		15	0.055			
		25	0.050			
		34	0.055			
		45	0.080			
Total		22	0.084	melt grown	300	
		22	0.047	hot pressed		
		25	0.058	hot pressed		

PROPERTY	SYMBOL	VALUE			UNIT	NOTES	TEMP.(°K)	REFERENCES
Dielectric Constant		x	ϵ_∞	ϵ_0				
Optical	ϵ_∞							
Static	ϵ_0	25	12.1	13.6		vacuum evaporated, single	77	Potter &
		55	11.1	12.7		phase films, spectral		Stierwalt
		65	10.3	12.1		thermal emittance meas.		
Electrical	ρ		ρ					
Resistivity		22	4×10^{-4}		ohm-cm	melt grown $n_p \sim 10^{20}$	300	Sahm & Pruss
		22	10^{-3}			hot pressed		
		25	10^{-3}			hot pressed		
		85	3.5×10^{-3}			liquid epitaxy, films	100	Kozlov et al.
			3.0×10^{-3}			on GaAs, 80-100μ thick $n_n = 10^{18}$	300	
Seebeck Coeff.	Q		Q					
		22	73.5		$\mu V/°K$	melt grown	300	Sahm & Pruss
			76			hot pressed		
Mobility			μ_p					
Hole	μ_p	25	70		cm^2/V sec	hot pressed	300	Sahm & Pruss
		22	145			melt grown		
Electron	μ_n		μ_n					
		85	1700			epitaxial films	100,300	Kozlov et al.

μ_p	x	77°K	300°K	$n_p (10^{16} cm^{-3})(300°K)$			
	20	–	21	4.6	vapor grown		Clough &
	25	–	48	10	single crystal		Tietjen
	30	–	251	3.4	epitaxial films		
	38	15	20	900			
	69	30	24	150			
	79	80	45	29			
	90	160	55	120			
	96	–	400	0.3			
	98	700	190	1.3			
	99	400	220	0.18			
μ_n				$n_n (10^{16} cm^{-3})(300°K)$			
	96.7	–	5370	1.1			
	97.5	5550	2920	4.5			
	99	4975	3840	4.7			
μ_p	7		400		melt grown,	300	Thomas et al.
	23		200		homogeneous, polycrystalline $n_p \sim 10^{17} cm^{-3}$		

PROPERTY	SYMBOL	VALUE		UNIT	NOTES	TEMP.(°K)	REFERENCES
Effective Mass	m_n	x	m_n				
		85	0.11	m_0	epitaxial films	100	Kozlov et al.
			0.12			300	
Energy Gap	E_g		E_g				
		0	0.70		GaSb single crystal, optical meas.	300	Cardona

PROPERTY	SYMBOL	VALUE		UNIT	NOTES	TEMP.(°K)	REFERENCES
Energy Gap	E_g	x	E_g				
		0	0.720	eV	GaSb	300	Thomas et al.
		2	0.715		polycrystalline, melt		
		10	0.700		grown and zone refined		
		20	0.680(min)		$n_n = 10^{16}$		
		25	0.681				
		35	0.710		$n_p = 10^{17}$		
		85	1.130				
		90	1.200		optical meas.		
		95	1.300				
		100	1.420		GaAs		
		100	1.4257		single crystal, optical meas.	300	Zvara
		85	1.24		liquid phase epitaxy optical transmission meas.	300	Kozlov et al.
		75	1.08		liquid epitaxy films 2-15μ	300	Antypas & James
		85	1.18		thick, photoemission meas.		
		95	1.33				

			100°K	210°K	300°K			
		8	0.74	0.68	0.62	photoconductivity meas. on		Taylor & Fortin
		23	0.74	0.68	0.62	polycrystalline wafers, high-As, $n_n = 10^{16}$,		
		86	1.25	1.24	1.20	high-Sb, $n_p = 10^{17}$		

PROPERTY		x	E_o	E_1	$E_1 + \Delta_1$	E_2	NOTES	TEMP.(°K)	REFERENCES
Energy Band		0	0.70	2.08	2.48	4.21	GaSb optical meas. on single cr.	300	Cardona, Zallen & Paul
		0	0.625	2.18	2.50	4.25	optical meas., flash	300	Sirota & Matyas
		15	0.625	2.18	2.50	4.25	evaporated films, 0.3-0.8μ		
		20	0.630	2.25	2.50	4.25	thick, also bulk crystal-		
		35	0.635	2.30	2.55	4.30	line samples		
		45	0.85	2.32	2.55	4.34	E_g for films $> E_g$ for bulk		
		60	0.95	2.38	2.64	4.35			
		72.5	1.00	2.50	2.75	4.50			
		82.5	1.12	2.50	2.85	4.50			
		100	1.45	2.75	3.05	5.00	GaAs		
		100	1.429	2.904	3.13	4.99	single crystals, optical meas.	300	Williams & Rehn, Shaklee et al., Cardona et al.

PROPERTY	SYMBOL	x	TO	UNIT	NOTES	TEMP.(°K)	REFERENCES
Phonon Branch Spectra Transverse Optic	TO	25	31	meV	vacuum evaporated, single	77	Potter & Stierwalt
		55	32		phase films, spectral		
		65	32.8		thermal emittance meas.		
		7	29.5		infrared reflectivity,	300	Lucovsky & Chen
		11	29.7		polycrystalline wafers		
		91	33.1				
		93	33.2				

PROPERTY	SYMBOL		$D*$ ($cm/Hz^{0.5}/W$)	Peak Wavelength (μ)	NOTES	TEMP.(°K)	REFERENCES
Detectivity	$D*$	0	5.4×10^5	1.75	polycrystalline wafers	300	Taylor & Fortin
		8	5.8×10^5	1.87			
		23	2.2×10^5	1.87			
		86	1.6×10^6	1.0			
Photoemission Yield		0.1 to 0.2% at 1.06μ			graded bandgap photo-cathode ∿700μ A/lm sensitivity	300	Antypas & James

ANTYPAS, G.A. and L.W. JAMES. Liquid Epitaxial Growth of Gallium Arsenic Antimonide and Its Use as a High-Efficiency, Long-Wavelength Threshold Photoemitter. J. OF APPLIED PHYS., v. 41, no. 5, Apr. 1970. p. 2165-2171.

ANTYPAS, G.A. et al. Operation of III-V Semiconductor Photocathodes in the Semitransparent Mode. J. OF APPLIED PHYS., v. 41, no. 7, June 1970. p. 2888-2894.

BEDNAR, J. and K. SMIROUS. The Melting Point of Gallium and Indium Antimonide (In Ger.). CZECH, J. PHYS., v. 5, no. 4, 1955. p. 546.

CARDONA, M. Fundamental Reflectivity Spectrum of Semiconductors with Zincblende Structure. J. OF APPLIED PHYS., Suppl. to v. 32, no. 10, Oct. 1961. p. 2151-2155.

CARDONA, M. et al. Electroreflectance at a Semiconductor-Electrolyte Interface. PHYS. REV., v. 154, no. 3, Feb. 1967. p. 696-720.

CLOUGH, R.B. and J.J. TIETJEN. Vapor-Phase Growth of Epitaxial Gallium Arsenic Antimonide Using Arsine and Stibine. AIME METALL. SOC., TRANS., v. 245, no. 3, Mar. 1969. p. 583-585.

FOSTER, L.M. and J.F. WOODS. Thermodynamic Analysis of the III-V Alloy Semiconductor Phase Diagrams. ELECTROCHEM. SOC., J., v. 119, no. 4, Apr. 1972. p. 504-507.

KOZLOV, Yu.M. et al. Epitaxial Films of Gallium Antimonide-Gallium Arsenide Solid Solutions and Some of their Electrical and Optical Properties. SOVIET PHYS. SEMICONDUCTORS, v. 4, no. 9, Mar. 1971. p. 1571-1572.

LE GUILLOU, G. and H.J. ALBANY. Contributions by Longitudinal and Transverse Phonons to the Lattice Thermal Conductivity in Gallium Antimonide at Low Temperatures (In Fr.). J. DE PHYSIQUE., v. 31, no. 5/6, May/June 1970.

LUCOVSKY, G. and M.F. CHEN. Long Wave Optical Phonons in the Alloy Systems: Gallium Indium Arsenide, Gallium Arsenic Antimonide and Indium Arsenic Antimonide. SOLID STATE COMM., v. 8, no. 17, Sept. 1970. p. 1397-1401.

McSKIMIN, H.J. et al. Elastic Moduli of Gallium Antimonide Under Pressure and the Evaluation of Compression to 80 kbar. J. OF APPLIED PHYS., v. 39, no. 9, Aug. 1968. p. 4127-4128.

POTTER, R.F. and D.L. STIERWALT. Reststrahlen Frequencies for Mixed Gallium Arsenic Antimonide System. PHYS. OF SEMICONDUCTORS, PROC. OF THE 7th INTERNATIONAL CONF., ACAD. PRESS, N.Y. AND LONDON; PARIS 1964. p. 1111-1114.

REDIKER, R.H. et al. Electrical and Electro-Optical Properties of Interface-Alloy Heterojunctions. AIME METALL. SOC. TRANS., v. 223, no. 3, 1965. p. 463-467.

RICHMAN, D. Dissociation Pressures of Gallium Arsenide, Gallium Phosphide and Indium Phosphide and the Nature of III-V Melts. J. OF PHYS. AND CHEM. OF SOLIDS, v. 24, no. 9, Sept. 1963. p. 1131-1139.

SAHM, P.R. and T.V. PRUSS. Pressure-Sintered Gallium Antimonide-Gallium Arsenide Alloys-Densification and Thermoelectric Properties. AIME METALL. SOC., TRANS., v. 239, no. 10, Oct. 1967. p. 1523-1526.

SHAKLEE, K.L. et al. Electroreflectance and Spin-Orbit Splitting in III-V Semiconductors. PHYS. REV. LETTERS, v. 16, no. 2, Jan. 1966. p. 48-50.

SIROTA, N.N. and E.E. MATYAS. Reflection and Absorption Spectra of Gallium Arsenic Antimonide Solid Solutions. PHYS. STATUS SOLIDI, v. 4, no. 2, Feb. 1971. p. K143-K146.

STRAUMANIS, M.E. and C.D. KIM. Solid Solubility in the System Gallium Antimonide-Gallium Arsenide. ELECTROCHEM. SOC., J., v. 112, no. 1, Jan. 1965. p. 112-113.

TAYLOR, A.E. and E. FORTIN. Photoconductivity in Some III-V Alloys. CANADIAN J. OF PHYS., v. 48, no. 16, Aug. 1970. p. 1874-1878.

THOMAS, M.B. et al. Energy Gap Variation in Gallium Arsenic Antimonide Alloys. PHYS. STATUS SOLIDI, v. 2, no. 3, July 1970. p. K141-K143.

WILLIAMS, E.W. and V. REHN. Electroreflectance Studies of Indium Arsenide, Gallium Arsenide and Gallium Indium Arsenide Alloys. PHYS. REV., v. 172, no. 3, Aug. 1968. p. 798-810.

WOOLLEY, J.C. Solid Solution of III-V Compounds. In: COMPOUND SEMICONDUCTORS; Preparation of III-V Compounds, v. 1, p. 3-20. Edited by: WILLARDSON, R.K. and H.L. GOERING, Reinhold Pub. Corp., N.Y. Chapman and Hall, Ltd., London.

ZALLEN, R. and W. PAUL. Effect of Pressure on Interband Reflectivity Spectra of Germanium and Related Semiconductors. PHYS. REV., v. 155, no. 3, Mar. 1967. p. 703-711.

ZVARA, M. Faraday Rotation and Faraday Ellipticity in the Exciton Absorption Region of Gallium Arsenide. PHYSICA STATUS SOLIDI, v. 36, no. 2, Dec. 1969. p. 785-792.

GALLIUM-ARSENIC-PHOSPHORUS SYSTEM

PROPERTY	SYMBOL	VALUE	UNIT	NOTES	TEMP.(°K)	REFERENCES

Formula GaP_xAs_{1-x}

Density

x	G*	G** (gr/cm^3)
0	–	5.32
13	5.20	–
38	4.89	
56	4.66	
60	4.62	
66	4.57	
72	4.51	
74	–	4.48
78	–	4.42
92	–	4.23-4.36
100	4.14	4.16

NOTES: GaAs (row x=0); GaP (row x=100)
TEMP.(°K): 293
REFERENCES: *Jones et al. **Abagyan et al.

Color: yellow to dark cherry red
NOTES: transparent, vapor transport preparation of elongated tablets
REFERENCES: Abagyan et al.

Lattice Parameters a_o

x	Jones	Rubenstein	Cooper	Pierron et al.	Abagyan et al.	
	vapor epitaxy single cr.	single cr. I-vapor transport		I-vapor transport	gas-transport single cr.	
0	5.65332 (calc.)	5.6532	5.64191		5.6527	GaAs
10		5.6305				
13	5.618					
20		5.6103				
30		5.5890				
38	5.578					
40		5.5676				
41				5.5687		
50		5.5483		5.5565		
55					5.5624	
56	5.550					
60	5.538	5.5296				
66	5.510					
70		5.5094				
72	5.503					
74					5.501	
78					5.499	
80		5.4894				
90		5.4704				
92					5.473-5.482	
100	5.4505	5.4508	5.4495		5.4505	GaP

Melting Point M.P.

x	M.P.			
0	1238	°C	GaAs	Richman
5	1245			Osamura & Murakami, Antypas, Osamura et al.
15	1270			
45	1325			
60	1350			
65	1360			
73	1380			
90	1420			
100	1467		GaP	Richman

Thermal Expansion Coeff.

x	value	unit	notes	references
0	6.96	10^{-6}/°K	GaAs	Pierron et al.
41	5.41		from lattice constant	
50	5.91		measurements	
100	5.81		GaP	

PROPERTY	SYMBOL	VALUE	UNIT	NOTES	TEMP.(°K)	REFERENCES

Thermal Conductivity — k

x	k (W/cm °K) 30°K	k (W/cm °K) 273°K
10	1	0.22
20	–	0.20
33–35	0.4	0.18
50	0.22	0.14

NOTES: polycrystalline, Te- Se- and Si-doped $n_n = 2\text{-}4 \times 10^{18}$

REFERENCES: Carlson et al.

Dielectric Constant Optical — ε_∞

x	ε_∞ 87°K	ε_∞ 300°K
0	10.4944	10.7479
6	10.3714	10.6314
12.5	10.2313	10.4601
25	10.0866	10.2909
35	9.9871	10.1632
41.7	9.8382	10.0331
62.5	9.5607	9.6203
100	8.8599	8.9980

NOTES: GaAs single crystals grown by closed tube, iodine vapor transport method; optical meas. in infrared. GaP

REFERENCES: Clark & Holonyak

x	300°K
6	10.76
28	10.20
56	9.53
85	8.86
99	8.45

NOTES: optical reflectivity meas. $n < 10^{16}$ cm^{-3}

REFERENCES: Verleur & Barker

Mobility Electron — μ_n

x	μ_n (cm^2/V sec)
38	3150

NOTES: Gas transport, vapor phase, epitaxial, single crystals, $n_n = 1.5 \times 10^{15}$

TEMP.: 300

REFERENCES: Ogirima & Kurata

x	μ_n 77°K	μ_n 300°K
0–30	15000	5000
40		1300
70		500

NOTES: epitaxial vapor deposition on (100) GaAs, not doped, $n_n = 5 \times 10^{15}\text{-}10^{16}$

REFERENCES: Tietjen & Weisberg

x	$n = 1.5 \times 10^{17}$	$n = 6 \times 10^{17}$
0 GaAs	5000	4000
5	–	4000
10	5000	4000
15	4000	3000
25	2500	–
30	1500	800
50	250	–

NOTES: epitaxial, n-type, single crystal, closed tube, vapor deposition on (110) GaAs or GaP, iodine transport, Se, Te or Sn-doped

TEMP.: 300

REFERENCES: Ku

x	μ_n	n_n (10^{18}cm^{-3})
12	1580	1.0
20	900	–
25	700	1.0
34	400	–
45	100	0.3
50–60	25	1.0

NOTES: epitaxial, n-type, single crystal, vapor deposition, Se and Te doped

TEMP.: 300

REFERENCES: Wolfe et al.

x	μ_n (cm^2/V sec) 125°K	μ_n (cm^2/V sec) 300°K	Dopant	n_n (10^{18}cm^{-3})
70	160	85	Te	1.8
75	140	90	Te	2.1
80	80	60	Te+Zn	1.3
80	100	70	Te	5.5
80	150	90	Se	1.4
90	300	100	–	0.4

NOTES: single crystals

REFERENCES: Yurova et al.

PROPERTY	SYMBOL	VALUE	UNIT	NOTES	TEMP.(°K)	REFERENCES

Effective Mass

Electron — m_n

x	m_n
14	0.085 m_o

300 — Hill, Craford et al.

Composition Coeff. — $m_n = 0.072\,(1+x)$

x	$m_n(m_o)$	$n_n(10^{18}\,cm^{-3})$
20	0.12	2.7–4.1
25	0.15	5.7
30	0.18	2.5
55	0.47	2.5
72	0.47	2.5

optical reflectivity and Faraday rotation at 2-24µ on n-type material — 300 — Iglitsyn et al.

Energy Band Structure

Direct Gap, E_o

Spin-Orbit Splitting Δ_o

x	E_o	Δ_o	E_1	Δ_1	E_o'	E_2
0	1.43	0.33	2.90	0.23	4.46	4.99 eV
10	1.55	0.32	2.94	0.24	4.48	5.01
20	1.67	0.28	3.01	0.22	4.52	5.03
30	1.82	0.27	3.06	0.23	4.52	5.04
40	1.90	0.25	3.14	0.23	4.58	
50	2.04	0.22	3.21	0.21	4.61	5.13
60	2.16	0.19	3.27	0.19	4.63	5.17
70	2.29	0.18	3.38		4.67	5.21
80	2.44	0.16	3.41			
90	2.60	0.12	3.58		4.75	5.24
100	2.75	0.09	3.66		4.75	5.28

GaAs electroreflectivity meas., sealed tube, iodine transport, polycrystalline, $n = 10^{17}$, GaAs and GaP are single crystals. GaP — 300 — Thompson et al., Irzikevicius et al.

x	E_o	Δ_o	E_1	Δ_1	E_o'
20	1.645	0.330	3.003	0.232	4.61

electroreflectance meas. — 300 — Rehn

x	$E_o + \Delta_o$	E_1
20	2.021	3.053

180

x	Δ_o
28	0.275
43	0.242
47	0.230
70	0.190
78	0.170
87	0.150

optical meas. on single crystal, epitaxial films — 300 — Hodby, Belle et al.

Indirect Gap — E_i

x	E_i(eV) Ku	E_i(eV) Spitzer & Mead
25	1.85	
30	1.90	
40		1.85
50	2.0	
55		1.92
75	2.12	
80		2.06
85		2.12
100		2.19

optical meas. on I-vapor deposited, single crystal, epitaxial films — 300 — Ku

photovoltaic and luminescence meas. on polycrystals — Spitzer & Mead

x	E_i(eV)
43–44	2.06 (cross over)
45	1.96 (cross over)

open tube, vapor deposited epitaxial film diodes on (100) GaAs, 5µ junction depth; electroluminescence meas. — 77, 300 — Herzog et al.

GALLIUM-ARSENIC-PHOSPHORUS SYSTEM

PROPERTY	SYMBOL	VALUE	UNIT	NOTES	TEMP.(°K)	REFERENCES

Energy Band Structure

x		E_1		$E_1+\Delta_1$		E_0'		E_2 (eV)		NOTES	REFERENCES
		80°K	295°K	80°K	295°K	80°K	295°K	80°K	295°K		
0 GaAs		3.0	2.9	3.22	3.13	4.45	4.42	5.1	5.05	sealed tube,	Woolley et al.,
10		3.09	2.95	3.24	3.15	4.53	4.5	5.15	-	iodine transport	Bergstresser
20		3.15	3.02	3.35	3.22	4.60	4.55	5.18	5.1	polycrystalline	et al.
30		3.20	3.10	3.38	3.30	4.60	4.5	5.20	5.25	or epitaxial	
40		3.25	3.18	3.39	3.32	4.65	4.60	-	-	layers, reflec-	
50		3.35	3.25	3.50	3.43	4.68	4.60	5.30	5.18	tivity meas.	
60		3.40	3.35	3.59	3.50	-	-	-	-		
70		3.52	3.42	-	-	4.77	-	5.32	-	reflectivity meas.	Williams &
80		-	3.50	-	-	-	-	-	-	on epitaxial layers	Jones
100 GaP		3.78	3.70	-	-	4.89	4.78	5.41	5.30		

PROPERTY	SYMBOL	VALUE	TEMP.(°K)	REFERENCES
Temperature Coeff.	dE_1/dT	$\approx -5 \times 10^{-4}$ eV/°K	80-295	Woolley et al.

Energy Gap E_0

x	77°K	295°K	dE_0/dT (10^{-4}eV/°K)	NOTES	REFERENCES
16	1.73	1.62	3.2	optical meas. on	Subashiev &
29	1.84	1.77	3.64	epitaxial, 2μ layers,	Chalikyan
41	2.02	1.94	3.64	n~10^{17}	
52	2.18	2.10	3.64		
70.5	2.44	2.36	3.64		
84	2.64	2.56	3.64		

PROPERTY	VALUE	TEMP.(°K)	REFERENCES
Composition Coeff.	$E_0 = 1.51 + 1.16x + 0.2x^2$	77	Subashiev &
	$= 1.43 + 1.15x + 0.2x^2$	295	Chalikyan
Temperature Coeff.	$dE_0/dT \quad \approx -5 \times 10^{-4}$ eV/°K	77-295	Subashiev & Chalikyan

Pressure Coeff. dE_g/dP

x	Value	NOTES	TEMP.(°K)	REFERENCES
40	1.3×10^{-5}/kg cm^{-2}	polycrystalline, Te-doped P to 10 kbars; electrical meas.	77	Likhter & Pel

Energy Levels
Donor E_d
Acceptor E_a

x	Dopant	E_d	E_a (eV)	dE_d/dP (eV/kbar)	NOTES	TEMP.(°K)	REFERENCES
37.5	Te	0.03		10.5×10^{-3}	electrical meas.	55-400	Craford et al.
30	S	0.04					
45	S	0.21		10.8×10^{-3}			
30	S	0.06		10.0×10^{-3}		77	
50	Fe		0.61		thermally stimulated conductivity, 2×10^{15}	90-350	Schade

Phonon Branch Spectra
Transverse Optic TO

x	TO	UNIT	NOTES	TEMP.(°K)	REFERENCES
6	44.6	meV	reflectivity meas.	300	Verleur &
28	46.5		n~10^{16}		Barker
56	47.8				
85	49.8				
100	50.2		GaP		

x	TO	LO ^{L}	LA	TA	TO	LO ^{X}	LA	TA	NOTES	TEMP.(°K)	REFERENCES
5.4	43.8			10.4					open tube	300	Chen et al.
17.7	44.0		32.8	10.4					vapor trans-		
35	44.1		33.0	10.4					port epitaxial		
48	44.4		33.4	10.2					films on (111)		
67.5	45.0		33.9	9.55					GaAs		
85	45.7	39.8	34.5	8.80	44.4	37.2	33.8	14.9			
100	46.8	40.8	34.7	8.49	44.7	38.3	32.8	14.5			

GALLIUM-ARSENIC-PHOSPHORUS SYSTEM

PROPERTY	SYMBOL	VALUE	UNIT	NOTES	TEMP.(°K)	REFERENCES

Magnetic Susceptibility — Symbol χ mol

x	$-\chi$ mol			
0	32	10^{-6} cgs	GaAs	300
30	30		single crystal, $n \sim 10^{17}$	
45	28		Faraday rotation meas.	
71	28		at 77–300°K	
75	27			
100	27		GaP	

References: Andrianov et al.

Refractive Index — Symbol x

x	87°K Wavelength				300°K Wavelength				Notes	References
	2.07μ	1.03μ	0.78μ	0.62μ	2.07μ	1.03μ	0.78μ	0.62μ		
0	3.27	3.43			3.33	3.47			closed tube,	Clark &
6	3.26	3.4			3.30	3.45			halogen vapor	Holonyak
12.5	3.22	3.35			3.26	3.41			transport,	
25	3.21	3.31	3.45		3.24	3.35	3.52		polycrystals,	
35	3.18	3.28	3.40		3.22	3.33	3.47		Se- or Te-doped,	
41.7	3.16	3.25	3.37		3.19	3.29	3.43		$n \sim 10^{18}$	
62.5	3.10	3.20	3.30	3.47	3.14	3.23	3.35	3.52		
100	–	3.07	3.15	3.28	–	3.11	3.20	3.33		

Photoemission Quantum Yield — Symbol Y
*(Electrons/quantum)

x	Y*	Wavelength (Å)	Notes	TEMP.(°K)	References
25	38 mA/W	4000	cesium activated p-type photocathode	300	Simon et al.
0–17	0.21*	6000	cesium coated, Zn-doped,		Garbe
30	0.10	6000	closed tube, iodine		
	0.25	5000	vapor transport crystals		
70	0.01	5000			
	0.20	4100			
100	0.19	4100			
	0.0	5000			

Diode Properties
Brightness B
Quantum Efficiency η
Current Density CD

x	B(fL)	Wavelength (Å)	η (%)	CD (A/cm²)	Notes	TEMP.(°K)	References
40	720	6520	0.2	4.4	Zn-diffused, vapor	300	Herzog et al.
29			0.6	4.4	grown, epitaxial films, $n \sim 10^{16}$–10^{17}		
38		9500	0.2	0.08	2×10^{16}	300	Ogirima & Kurata
40	1000	6450		10	vapor grown, 10^{17}–10^{18}	300	Burmeister et al.
40	max.	6530	0.06	10	$n \sim 10^{17}$, selenium doped epitaxial layers, cathodoluminescence meas. 1.2–1.5μ junction depth	300	Heath & Stewart
76	200	5850	0.002	16	Zn-diffused, vapor grown, epitaxial diode, 5.5×10^{16}, 0.3–0.4 mm² area	296	Epstein & Huebner
45	400–600 (at 10 A/cm²)	6840	0.5 / 0.2	20	Zn-N doped Zn-doped, vapor phase, epitaxial EL diodes	300	Groves et al.
34–37			2.75 / 0.21		EL diodes $n \sim 4 \times 10^{17}$	78 / 300	Maruska & Pankove
38	>300	6800	0.035	10	vapor grown EL diodes	300	Nuese et al. (A, B)
42	8500	6600	0.4			77	

PROPERTY	SYMBOL	VALUE				NOTES	TEMP.($^\circ$K)	REFERENCES
		x	$\overset{\circ}{A}$	η (%)	TCD (A/cm^2)			
Laser Properties Threshold Current Density	TCD	20.0	7250		9×10^2	vapor deposited,	78	Tietjen et al.
		40.5	6750		9×10^5	epitaxial films	300	
		14	8100	26	9×10^2	25W power output	300	
		10	7850		$1.1\text{-}1.3 \times 10^3$	Te-doped, vapor grown	77	Eliseev et al.
		15	7585			single crystal, epi-		
		20	7400			taxial films 10-16μ		
		30	6890			junction depth		
		35	6640					
		45	6390					
		30	6750			vapor grown, thin platelets	77	Johnson & Holonyak
Memory Effect, (Lifetime)		3 millisec. at 6500 $\overset{\circ}{A}$					77	Eliseev & Ismailov

ABAGYAN, S.A. et al. X-Ray and Optical Investigations of Gallium Arsenic Phosphide Crystals. SOVIET PHYS. SOLID STATE, v. 7, no. 1, July 1965. p. 153-157.

ALLEN, J.W. et al. Microwave Oscillations in Gallium Arsenic Phosphorus Alloys. APPLIED PHYS. LETTERS, v. 7, no. 4, Aug. 15, 1965. p. 78-80.

ANDRIANOV, D.G. et al. Magnetic Susceptibility of Solid Solutions in the Gallium Arsenide-Gallium Phosphide System. SOVIET PHYS. SEMICONDUCTORS, v. 4, no. 8, Feb. 1971. p. 1268-1270.

ANTYPAS, G.A. The Ga-GaP-GaAs Ternary Phase Diagram. ELECTROCHEM. SOC., J., v. 117, no. 5, May 1970. p. 700-703.

BAN, V.S. Mass Spectrometric Studies of Vapor Phase Crystal Growth. 1. $GaAs_xP_{1-x}$ System (0<x<1). ELECTRO-CHEM. SOC., J., v. 118, no. 9, Sept. 1971. p. 1473-1478.

BAN, V.S. et al. Influence of Deposition Temperature on Composition and Growth Rate of $GaAs_xP_{1-x}$ Layers. J. OF APPLIED PHYS., v. 43, no. 5, May 1972. p. 2471-2472.

BELLE, M.L. et al. Optical Reflection of Gallium Phosphide, Gallium Arsenide, and Their Solid Solutions. SOVIET PHYS. SOLID STATE, v. 8, no. 9, Mar. 1967. p. 2098-2101.

BERGSTRESSER, T.K. et al. Reflectivity and Band Structure of Gallium Arsenide, Gallium Phosphide, and Gallium Arsenic, Phosphorus Alloys. PHYS. REV. LETTERS, v. 15, no. 16, Oct. 18, 1965. p. 662-664.

BURMEISTER, R.A. JR., et al. Large Area Epitaxial Growth of Gallium Arsenic Phosphide for Display Applications. AIME METALL. SOC., TRANS., v. 245, no. 3, Mar. 1969. p. 587-592.

CARLSON, R.O. et al. Thermal Conductivity of Gallium Arsenide and Gallium Arsenic Phosphides Laser Semiconductors. J. OF APPLIED PHYS., v. 36, no. 2, Feb. 1965. p. 505-507.

CHEN, Y.S. et al. Lattice Vibration Spectra of Gallium Arsenic Phosphide Single Crystals. PHYS. REV., v. 151, no. 2, Nov. 11, 1966. p. 648-656.

CLARK, D. JR. and N. HOLONYAK, JR. Optical Properties of Gallium Arsenide-Phosphide. PHYS. REV., v. 156, no. 3, Apr. 15, 1967. p. 913-924.

COOPER, A.S. Precise Lattice Constants of Germanium, Aluminum, Gallium Arsenide, Uranium, Sulfur, Quartz and Sapphire. ACTA CRYSTALLOGRAPHICA, v. 15, 1962. p. 578-582.

CRAFORD, M.G. et al. Effect of Tellurium and Sulfur Donor Levels on the Properties of Gallium Arsenic Phosphide Near the Direct-Indirect Transition. PHYS. REV., v. 168, no. 3, Apr. 15, 1968. p. 867-882.

DEAN, P.J. et al. Low-Level Interband Absorption in Phosphorus-Rich Gallium Arsenide-Phosphide. PHYS. REV., v. 181, no. 3, May 15, 1969. p. 1149-1153.

ELISEEV, P.G. et al. Gallium Phosphorus Arsenic-Based Injection Lasers. SOVIET PHYS. SEMICONDUCTORS, v. 2, no. 4, Oct. 1968. p. 507-508.

ELISEEV, P.G. and I. ISMAILOV. Memory Effect in Injection Lasers. SOVIET PHYS. TECH. PHYS., v. 13, no. 12, June 1969. p. 1671-1672.

EPSTEIN, A.S. and R.C. HUEBNER. Yellow Emission from Gallium Arsenic Phosphide Diodes. SOLID STATE ELECTRONICS, v. 12, no. 6, June 1969. p. 494-496.

FENNER, G.E. Pressure Effect on Resistivity of Gallium Arsenic Phosphides. PHYS. REV., v. 134, no. 4A, May 18, 1964. p. A1113-A1118.

GARBE, S. Photoemission from Gallium Arsenic Phosphide Covered with Low Work Function Layers. PHYS. STATUS SOLIDI, v. 33, no. 2, June 1969. p. K87-K91.

GROVES, W.O. et al. The Effect of Nitrogen Doping on Gallium Arsenic Phosphide Electroluminescent Diodes. APPLIED PHYS. LETTERS, v. 19, no. 6, Sept. 15, 1971. p. 184-186.

HERZOG, A.H. et al. Electroluminescence of Diffused Gallium Arsenic Phosphide Diodes with Low Donor Concentrations. J. OF APPLIED PHYS., v. 40, no. 4, Mar. 15, 1969. p. 1830-1838.

HILL, D.E. Effective Mass of the (000) Conduction Band of $GaAs_{1-x}P_x$. AMERICAN PHYS. SOC., BULL., v. 11, Ser. 2, Mar. 1966. p. 205.

HODBY, J.W. Infra-red Absorption in Gallium Phosphide-Gallium Arsenide Alloys. II. Absorption in p-Type Material. PHYS. SOC., PROC., Pt. 2, v. 82, no. 526, Aug. 1963. p. 324-326.

IGLITSYN, M.I. et al. Some Features of the Structure of the Conduction Band of Gallium Arsenic Phosphide Solid Solutions in the Intermediate Range of Compositions. SOVIET PHYS. SEMICONDUCTORS, v. 3, no. 12, June 1970. p. 1509-1513.

IRZIKEVICIUS, A. et al. The Investigation of the Electroreflectance Spectra of Gallium Arsenic Phosphide (In Russ.). LIETUVOS FIZ. RINKINYS, v. 9, no. 3, 1969. p. 535-545.

JOHNSON, M.R. and N. HOLONYAK, JR. Optically Pumped Thin-Platelet Semiconductor Lasers. J. OF APPLIED PHYS., v. 39, no. 8, July 1968. p. 3977-3985.

KU, S.M. The Preparation and Properties of Vapor-Grown Gallium Arsenide-Gallium Phosphide Alloys. ELECTRO-CHEM. SOC., J., v. 110, no. 9, Sept. 1963. p. 991-995.

LIKHTER, A.I. and E.G. PEL. Investigation of $GaAs_{0.6}P_{0.4}$ at Pressures up to 10 kbars. SOVIET PHYS. SEMICON-DUCTORS, v. 5, no. 9, Mar. 1972. p. 1508-1510.

LOGAW, R.A. et al. Electroluminescence in $GaAs_xP_{1-x}$, $In_xGa_{1-x}P$ and $Al_xGa_{1-x}P$ Junctions with $x \lesssim 0.01$. J. OF APPLIED PHYS., v. 42, no. 6, May 1971. p. 2328-2335.

MARUSKA, H.P. and J.I. PANKOVE. Efficiency of Gallium Arsenic Phosphide Electroluminescent Diodes. SOLID STATE ELECTRONICS, v. 10, no. 9, Sept. 1967. p. 917-925.

MONSANTO CO., ST. LOUIS, MO. CENTRAL RES. DEPT. Manufacturing Methods for Epitaxially Growing Gallium Arsenide-Gallium Phosphide Single Crystal Alloys. Interim Eng. PR, Sept. 1, 1967-Nov. 30, 1967. Contract no. AF 33-615-3618. Nov. 1967. 51 p.

NEAMEN, D.A. and W.W. GRANNEMANN. Electrical Characteristics of GaAsP Schottky Barrier Diodes. SOLID STATE ELECTRONICS, v. 14, 1971. p. 1319-1323.

NUESE, C.J. et al. Electroluminescence of Vapor-Grown Gallium Arsenide and Gallium Arsenide-Phosphide Diodes. AIME METALL. SOC., TRANS., v. 242, no. 3, Mar. 1968. p. 400-406.

NUESE, C.J. et al. Optimization of Electroluminescent Efficiencies for Vapor-Grown Gallium Arsenic Phosphide Diodes. ELECTROCHEM. SOC., J., v. 116, no. 2, Feb. 1969. p. 248-253.

OGIRIMA, M. and K. KURATA. Effect of Donor Concentration on Several Properties of Gallium Arsenide Phosphide. JAPAN. J. OF APPL. PHYS., v. 11, no. 3, Mar. 1972. p. 331-337.

OSAMURA, K. and Y. MURAKAMI. Phase Diagram of Gallium Arsenide-Gallium Phosphide Quasi-Binary System. JAPAN. J. OF APPL. PHYS., v. 8, no. 7, July 1969. p. 967.

OSAMURA, K. et al. Experiments and Calculation of the Gallium-Gallium Arsenide-Gallium Phosphide Ternary Phase Diagram. ELECTROCHEM. SOC., J., v. 119, no. 1, Jan. 1972. p. 103-108.

PANKOVE, J.I. Temperature Dependence of Emission Efficiency and Lasing Threshold in Laser Diodes. IEEE J. OF QUANTUM ELECTRONICS, v. QE-4, no. 4, Apr. 1968. p. 119-122.

PIERRON, E.D. et al. Coefficient of Expansion of Gallium Arsenide, Gallium Phosphide and Gallium Arsenic Phosphide Compounds from 62 to 200°C. J. OF APPLIED PHYS., v. 38, no. 12, Nov. 1967. p. 4669-4671.

REHN, V. Electroreflectance in $GaAs_xP_{1-x}$. AMERICAN PHYS. SOC., BULL., v. 11, Ser. 2, Mar. 1966. p. 205.

RICHMAN, D. Dissociation Pressure of Gallium Arsenide, Gallium Phosphide and Indium Phosphide and the Nature of III-V Melts. J. OF PHYS. AND CHEM. OF SOLIDS, v. 24, no. 9, Sept. 1963. p. 1131-1139.

RUBENSTEIN, M. The Preparation of Homogeneous and Reproducible Solid Solutions of Gallium Phosphide-Gallium Arsenide. ELECTROCHEM. SOC., J., v. 112, no. 4, Apr. 1965. p. 426-430.

SCHADE, H. Trapping Phenomena in Iron-Doped Gallium Arsenic Phosphide. HELV. PHYS. ACTA, v. 41, no. 6/7, 1968. p. 1142-1151.

SIMON, R.E. et al. Gallium Arsenic Phosphide as a New High Quantum Yield Photoemissive Material for the Visible Spectrum. APPLIED PHYS. LETTERS, v. 15, no. 2, July 15, 1969. p. 43.

SPITZER, W.G. and C.A. MEAD. Conduction Band Minima of Gallium Arsenide-Gallium Phosphide Systems. PHYS. REV., v. 133, no. 3A, Feb. 3, 1964. p. A872-A875.

SUBASHIEV, V.K. and G.A. CHALIKYAN. Direct Transitions and Spin-Orbit Splitting in Gallium Phosphorus Arsenide. SOVIET PHYS. SEMICONDUCTORS, v. 3, no. 10, Apr. 1970. p. 1216-1219.

THOMPSON, A.G. et al. Electroreflectance in the Gallium Arsenide-Gallium Phosphide Alloys. PHYS. REV., v. 146, no. 2, June 10, 1966. p. 601-610.

TIETJEN, J.J. et al. Vapor-Phase Growth of Gallium Arsenic Phosphide Room-Temperature Injection Lasers. AIME METALL. SOC., TRANS., v. 239, no. 3, Mar. 1967. p. 385-387.

VERLEUR, H.W. and A.S. BARKER, JR. Infrared Lattice Vibrations in Gallium Arsenic Phosphorus Alloys. PHYS. REV., v. 149, no. 2, Sept. 16, 1966. p. 715-729.

WILLIAMS, E.W. and C.E. JONES. Reflectivity Measurements on Epitaxial Gallium Arsenide-Gallium Phosphide Alloys. SOLID STATE COMMUNICATIONS, v. 3, no. 8, Aug. 1965. p. 195-198.

WOLFE, C.M. et al. Growth and Dislocation Structure of Single-Crystal Gallium Arsenide-Gallium Phosphide Systems. J. OF APPLIED PHYS., v. 36, no. 12, Dec. 1965. p. 3790-3801.

WOOLLEY, J.C. et al. Reflectivity of Gallium Arsenic Phosphide Alloys. PHYS. REV. LETTERS, v. 15, no. 16, Oct. 18, 1965. p. 670-672.

YURLOVA, E.S. et al. Electron Mobility in Gallium Arsenic Phosphide Solid Solutions. SOVIET PHYS. SEMICON-DUCTORS, v. 4, no. 8, Feb. 1971. p. 1346-1348.

JONES, V.O., V. REHN and D.S. KYSER. To be published.

GALLIUM-INDIUM-ANTIMONY SYSTEM

PROPERTY	SYMBOL	VALUE	UNIT	NOTES	TEMP.(°K)	REFERENCES
Formula		$Ga_xIn_{1-x}Sb$				
Symmetry		cubic				

Lattice Parameter a_o

x	a_o		NOTES	REFERENCES
0	6.47877	Å	InSb	Giesecke & Pfister
5	6.453		homogeneous, polycrystalline	Kolm et al., Kleshchinskii et al., Adler
10	6.43			
14	6.427			
20	6.394			
65	6.220			
73	6.187			
74	6.185			
85	6.1170			
90	6.1013			
100	6.094		GaSb	Donnay

Melting Point M.P.

x	M.P. (°C) Liquidus	Solidus	NOTES	REFERENCES
0	525.2		InSb	Bednar & Smirous
0	525		annealed, homogeneous	Woolley & Lees, Steininger, Stringfellow
10	544	531		
20	575	537		
30	600	544		
40	622	550		
50	644	562.5		
60	656	581		
70	675	588		
80	687	600		
90	700	650		
95	706	675		
100	711		GaSb	
100	712.1		GaSb	Bednar & Smirous

Thermal Conductivity k

x	k(W/cm °K)	NOTES	TEMP.(°K)	REFERENCES
0	0.15	InSb	300	Woolley & Briggs
10	0.075	polycrystalline, 1 mole% inhomogeneity		
20	0.06			
30	0.06			
40	0.053			
50	0.053			
55	0.053			
70	0.054			
80	0.068			
90	0.012			
100	0.024	GaSb		

	k(W/cm °K)					
x	4.2°K	10°K	40°K	300°K	NOTES	REFERENCES
67.5	0.25	0.44	0.32	0.07	polycrystalline	Briggs & Challis
87.5	0.52	0.75	0.45	0.08		
93.25	0.38	0.65	0.53	-		
95.5	0.25	0.53	0.80	-		
97.0	0.20	0.44	0.80	-		

x	300°K	700°K	NOTES	REFERENCES
33	0.056	0.037	polycrystalline, n- or p-type Te or Zn-doped	Kudman et al.
55	0.063	0.033		
82	0.090	0.042		

PROPERTY	SYMBOL	VALUE			UNIT	NOTES	TEMP.(°K)	REFERENCES
Dielectric Constant Optical	ε_∞	x	ε_∞			melt-grown, polycrystalline infrared reflectivity meas. at 35-80μ	300	Brodsky et al.
		5	17.5					
		15	15.7					
		30	15.1					
		50	14.9					
		60	17.5					
		70	16.0					
		84	17.0					
		95	17.0					

PROPERTY	SYMBOL	VALUE				NOTES	REFERENCES
Electrical Resistivity	ρ		$\rho(\Omega\text{-cm})$				
		x	100°K	300°K	500°K		Zeinalov & Aliev
		10	0.25	0.01	0.0016		
		25	0.10	0.03	0.005		
		50	0.05	0.10	0.010		
		75	0.1	0.20	0.063		
		90	0.1	0.25	0.17		
			$\rho(10^{-3}\Omega\text{-cm})$				
			100°K	300°K			
		94-96	0.02	4.5		single crystals	Averous et al.
		95.5	0.04	8.5			
		96.9	0.05	8.7			
		98.1	0.07	14.5			
		74-84	1.0			polycrystalline n- and p-type, 10^{17}-10^{20}	Kudman et al.

PROPERTY	SYMBOL	VALUE			NOTES	TEMP.(°K)	REFERENCES
Mobility Electron	μ_n		$\mu_n(\text{cm}^2/\text{V sec})$	$n_n(10^{17}\text{cm}^{-3})$			
		94	2.93×10^3	4.7	single crystals	300	Averous et al.
		96	3.55	9.7			
		96.9	1.0	2.4			
		35	12.6×10^3	14.0	polycrystalline	300	Kudman & Seidel
		55	12.5	6.3			
		66	8.8	12.9			
		83	9.7	2.8			
		0	6.0×10^4		InSb	300	Woolley & Gillett
		5	3.5		polycrystalline, $n_p = 8\times10^{16}$-4.5×10^{17} >40 mole% GaSb $n_n = 6\times10^{15}$-3×10^{16} <40 mole% GaSb		
		16	1.5				
		33	1.3				
		50	0.8				
		70	0.8				
		80	0.8				
		90	0.6				
		100	0.3		GaSb		
		50	6783	3.1	polycrystalline, Te-doped	300	Zeinalov et al.
			6470	5.6			
			4375	10.0			
			3310	57.0			
			1577	55.0			
Hole	μ_p	∿700			almost independent of composition	300	Ivanov-Omskii & Kolomiets

PROPERTY	SYMBOL	VALUE		UNIT	NOTES	TEMP.(°K)	REFERENCES
Seebeck Coeff.	Q	$\underline{x}$ $\quad$ $Q(\mu V/°K)$	$n(10^{17})$				
		35-83 $\quad$ -200	1-10		homogeneous, polycr.	300	Kudman & Seidel
		12-54 $\quad$ -(300-350)	3×10^{15}		homogeneous, polycr.	300	Title & Plasket
		50 $\quad\quad$ -183	3.1		Te-doped, polycr.	300	Zeinalov et al.
		-133	5.6				
		-116	10				
		- 44	57				
		0 $\quad\quad$ -238			InSb	300	Woolley &
		10 $\quad\quad$ -247			polycrystalline		Briggs
		44 $\quad\quad$ - 24					
		47 $\quad\quad$ +140					
		82 $\quad\quad$ +551					
		94 $\quad\quad$ +596					
		100 $\quad\quad$ + 39			GaSb		
		$\quad$ 300°K $\quad$ 600°K	800°K				
		84 $\quad$ 150 $\quad$ 200	250	$\mu V/°K$	polycrystalline, p-type		Kudman et al.

PROPERTY	SYMBOL	VALUE		UNIT	NOTES	TEMP.(°K)	REFERENCES
Nernst-Ettingshausen Coefficient		$\quad$ 100°K $\quad$ 340°K	400°K				
$\quad$ Transverse	$NE_{\perp}$	50 $\quad$ -0.5	-0.13	cgs	polycrystalline		Ivanov-Omskii
$\quad$ Longitudinal	$NE_{\parallel}$	-0.7 $\quad$ 0	+0.15				& Kolomiets
		$\quad$ 200°K	300°K				
Piezoresistance	π_{11}	94 $\quad$ 160	0	$10^{-11} cm^2/dyne$	single crystal		Averous et al.
	$\pi_{11}+2\pi_{12}$	1	1				

PROPERTY	SYMBOL	VALUE		UNIT	NOTES	TEMP.(°K)	REFERENCES
Effective Mass		x $\quad$ m_n					
$\quad$ Electron	m_n	12 $\quad$ 0.0142		m_o	macrocrystalline	300	Title &
$\quad$ Hole	m_p	22 $\quad$ 0.0162					Plaskett
		29 $\quad$ 0.0180					
		41 $\quad$ 0.0215					
		48 $\quad$ 0.0237					
		54 $\quad$ 0.0255					
			$n_n (10^{18} cm^{-3})$				
		50 $\quad$ 0.029	0.31		macrocrystalline	300	Zeinalov et al.
		0.033	0.56				
		0.039	1.00				
		0.058	5.70				
		0.056	5.50				
			m_p				
		0 $\quad$ 0.04	0.2		InSb	300	Woolley &
		10 $\quad$ 0.05	0.2		polycrystalline		Gillett
		28 $\quad$ 0.05	0.2		$n_p= 8 \times 10^{16} - 4.5 \times 10^{17}$		
		47 $\quad$ 0.06	0.25		$n_n= 6 \times 10^{15} - 3 \times 10^{16}$		
		60 $\quad$ 0.19	0.35				
		70 $\quad$ 0.12	0.40				
		90 $\quad$ 0.24	0.74				
		100 $\quad$ 0.35			GaSb		

x	m_n	n_n	m_p	n_p			
50	0.035	5×10^{17}	0.19	2.6×10^{17}	polycrystalline,	300	Gasanly &
	0.040	1.4×10^{18}	0.23	2.8×10^{18}	reflectivity meas.		Subashiev
	0.050	4.3×10^{18}	0.28	1.8×10^{19}	at 2-120μ		
	0.065	1.2×10^{19}	0.37	1.3×10^{20}			

30

GALLIUM-INDIUM-ANTIMONY SYSTEM

PROPERTY	SYMBOL	VALUE	UNIT	NOTES	TEMP.(°K)	REFERENCES

Energy Gap — E_g

x	E_g 4.2°K	E_g 300°K	UNIT	NOTES	REFERENCES
0	0.2355	0.180	eV	InSb magnetooptical meas.	Pidgeon & Brown, Zwerdling et al. (B)
0	0.236	–		InSb ESR and optical meas.	Title & Plaskett
12	0.249	0.188			
22	0.283	0.220			
29	0.313	0.248			
41	0.374	0.306			
48	0.412	0.342			
54	0.443	0.372			
100	0.813	–			
100	0.813	0.70		GaSb magnetoabsorption	Zwerdling et al. (A) Cardona, Lukes & Schmidt

Spin Orbit Splitting — Δ_o

x	E_g 0°K	E_g 300°K	Δ_o 0°K	Δ_o 300°K	UNIT	NOTES	REFERENCES
0	0.25	0.185	–	–	eV	InSb homogeneous, polycrystalline, $n = 10^{15}-10^{18}$, electrical meas., also optical meas. at 300°K	Coderre & Woolley, Woolley & Gillett, Woolley & Evans, Kolm et al., Ivanov-Omskii & Kolomiets
10	0.26	0.19	–	–			
20	0.30	0.23	–	–			
30	0.32	0.25	0.76	0.72			
45	0.37	0.31	0.78	0.72			
50	0.40	0.34	0.78	0.72			
60	0.48	0.40	0.80	0.74			
75	0.57	0.50	0.84	0.76			
80	0.65	0.55	0.85	0.77			
90	0.75	0.64	0.87	0.78			
100	0.82	0.75	0189	0.82		GaSb	

Band Structure

x	E_1	$E_1+\Delta_1$	E_2	UNIT	NOTES	TEMP.(°K)	REFERENCES
0	1.88	2.3	4.05	eV	InSb macrocrystalline, electroreflectance meas.	300	Vishnubhatla et al.
10	1.87	2.25	4.05				
20	1.85	2.25	4.05				
30	1.85	2.25	4.05				
40	1.86	2.25	4.07				
50	1.86	2.28	4.07				
60	1.89	2.3	4.07				
70	1.90	2.34	4.12				
80	1.95	2.34	4.18				
90	1.98	2.38	4.20				
100	2.0	2.45	4.25		GaSb		

Temperature Coeff.

x	dE_g/dT	UNIT	NOTES	TEMP.(°K)	REFERENCES
0-30	-2.75	10^{-4} eV/°K	optical meas. on polycrystalline samples	100-800	Woolley & Evans
65	-3.5				
75	-3.6				
100	-4.2				

GALLIUM-INDIUM-ANTIMONY SYSTEM

PROPERTY	SYMBOL	VALUE			UNIT	NOTES	TEMP.(°K)	REFERENCES
Phonon Branch Spectra		x	TO	LO				
Longitudinal Optic	LO							
Transverse Optic	TO	0	-	23.9	meV	InSb	300	Brodsky et al.
		5	24.9	25.3		polycrystalline		
		15	24.9	25.7				
		30	25.7	26.5				
		50	26.7	27.6				
		60	26.9	27.9				
		70	27.4	28.5				
		84	27.8	28.9				
		95	28.0	29.4				
		100	28.15	29.4		GaSb		
Magnetic Susceptibility	χ_{mol}	x	$-\chi_{mol}$					
		0	66		10^{-6} cgs	polycrystalline	300	Matyas
		20	56					
		40	50					
		84	40					
		100	38					

ADLER, P.N. Pressure-Induced Transformation Behavior of Some Quasi-Binary Alloys of InSb. J. OF PHYS. AND CHEM. OF SOLIDS, v. 30, no. 5, May 1969. p. 1077-1082.

AVEROUS, M. et al. Conduction Bands of GaSb-InSb Mixed Crystals with a Low Concentration of InSb. PHYS. STATUS SOLIDI, v. 29, no. 2, Oct. 1968. p. 807-812.

BAUKIN, I.S. et al. Concentration Dependence of the Forbidden Band Width in Alloys of InSb with GaSb. SOVIET PHYS.-SOLID STATE, v. 7, no. 4, Oct. 1965. p. 1019-1020.

BEDNAR, J. and K. SMIROUS. The Melting Point of Gallium and Indium Antimonide (In Ger.). CZECH, J. PHYS., v. 5, no. 4, 1955. p. 546.

BLAKEMORE, J.S. Properties of Gallium Indium Antimonide. CANADIAN J. OF PHYS., v. 35, no. 1, Jan. 1957. p. 91-97.

BRIGGS, A.G. and L.J. CHALLIS. Phonon Scattering by Acceptor Defects in GaSb and GaSb-InSb Alloys. J. OF PHYS., C, SER. 2, v. 2, no. 7, July 1969. p. 1353-1356.

BRIGGS, A.G. et al. The Thermal Conductivity of GaSb-InSb Alloys at 300 K. J. OF PHYS., C, SER. 2, v. 3, no. 3, Mar. 1970. p. 687-695.

BRODSKY, M.H. et al. Infrared Reflectivity Spectra of the Mixed Crystal System $Ga_{1-x}In_xSb$. PHYS. REV. B, SER. 3, v. 2, no. 8, Oct. 15, 1970. p. 3303-3311.

CAMPOS, M.D. et al. Sulphur Impurity State in $Ga_{1-x}In_xSb$. PHYS. STATUS SOLIDI, B, v. 47, no. 1, Sept. 1971. p. 137-141.

CARDONA, M. Fundamental Reflectivity Spectrum of Semiconductors with Zincblende Structure. J. OF APPLIED PHYS., suppl. to v. 32, no. 10, Oct. 1961. p. 2151-2155.

CODERRE, WM.M. and J. C. WOOLLEY. Conduction Bands of $Ga_xIn_{1-x}Sb$ Alloys. CANADIAN J. OF PHYS., v. 47, no. 22, Nov. 1969. p. 2553-2564.

DONNAY, J.D.H. (Ed.) Crystal Data. Determinative Tables, 2nd Ed. AMERICAN CRYST. ASSN., Apr. 1963. ACA Monograph no. 5.

GASANLY, N.M. and V.K. SUBASHIEV. Interaction of Plasmons with Optical Phonons in $In_{0.5}Ga_{0.5}Sb$. SOVIET PHYS.-SOLID STATE, v. 13, no. 1, July 1971. p. 124-127.

GASANLY, N.M. et al. Vibration Spectra of $In_xGa_{1-x}Sb$ Solid Solutions. SOVIET PHYS.-SOLID STATE, v. 13, no. 1, July 1971. p. 54-59.

GIESECKE, G. and H. PFISTER. Precision Determination of the Lattice Constants of III-V Compounds. ACTA CRYSTALLOGRAPHICA, v. 11, 1958. p. 369-371.

IVANOV-OMSKII, V.I. and B.T. KOLOMIETS. Charge Carrier Mobility and the Effective Mass of Electrons in Alloys of Indium and Gallium Antimonides. SOVIET PHYS.-SOLID STATE, v. 4, no. 1, July 1962. p. 216-218. (A)

IVANOV-OMSKII, V.I. and B.T. KOLOMIETS. The Dependence of the Energy Gap on Composition for the System InSb-GaSb (In Russ.). AKAD. NAUK SSSR. DOKLADY, v. 126, no. 6, July 1, 1959. p. 135-136. (B)

IVANOV-OMSKII, V.I. and B.T. KOLOMIETS. Thermomagnetic Effects in n-Type Gallium Antimonide and its Alloys with Indium Antimonide. SOVIET PHYS.-SOLID STATE, v. 3, no. 11, May 1962. p. 2581-2582. (C)

KLESHCHINSKII, L.I. et al. Thermal Conductivity of $In_{1-x}Ga_xSb$ Solid Solutions. SOVIET PHYS.-SEMICONDUCTORS, v. 2, no. 6, Dec. 1968. p. 675-677.

KOLM, C. et al. Studies on Group II-V Intermetallic Compounds. PHYS. REV., v. 108, no. 4, Nov. 15, 1957. p. 965-971.

KUDMAN, I. and T.E. SEIDEL. Conduction Bands in GaSb-InSb Alloys. J. OF APPLIED PHYS., v. 38, no. 11, Oct. 1967. p. 4379-4382.

KUDMAN, I. et al. High-Temperature Thermal and Electrical Properties of GaSb-InSb Alloys. J. OF APPLIED PHYS., v. 38, no. 12, Nov. 1967. p. 4641-4647.

LUKES, F. and E. SCHMIDT. The Fine Structure and the Temperature Dependence of the Reflectivity and Optical Constants of Germanium, Silicon and III-V Compounds. INTERNAT. CONF. ON THE PHYSICS OF SEMICONDUCTORS, PROC., Exeter, July 1962. Ed. by: A.C. STRICKLAND. London, Inst. of Phys. Soc., 1962. p. 389-394.

McGRODDY, J.C. et al. The Gunn Effect and Conduction Band Structure in $Ga_xIn_{1-x}Sb$ Alloys. SOLID STATE COMM., v. 7, no. 13, July 1969. p. 901-903.

MATYAS, M. The Magnetic Susceptibility of Solid Solutions of Semi-Conducting Compounds of $A^{III}B^V$. CZECH. J. OF PHYS., v. 11, no. 6, 1961. p. 461-463.

PIDGEON, C.R. and R.N. BROWN. Interband Magneto-Absorption and Faraday Rotation in Indium Antimonide. PHYS. REV., v. 146, no. 2, June 1966. p. 575-583.

ROBERT, J.-L. Study of Electron Diffusion Mode in Gallium Antimonide and Gallium Antimonide-Indium (In Fr.). SOLID STATE COMM., v. 7, no. 1, Jan. 1969. p. 143-147.

STEININGER, J. Thermodynamics and Calculation of the Liquidus-Solidus Gap in Homogeneous, Monotonic Alloy Systems. J. OF APPLIED PHYS., v. 41, no. 6, May 1970. p. 2713-2724.

STRINGFELLOW, G.B. Calculation of Ternary Phase Diagrams of III-V Systems. J. OF PHYS. AND CHEM. OF SOLIDS, v. 33, no. 3, Mar. 1972. p. 665-677.

TITLE, R.S. and T.S. PLASKETT. ESR Determination of Electron Effective Masses in $Ga_xIn_{1-x}Sb$ Alloys. J. OF APPLIED PHYS., v. 41, no. 1, Jan. 1970. p. 334-337.

VISHNUBHATLA, S.S. et al. Electroreflectance Measurements in Mixed II-V Alloys. CANADIAN J. OF PHYS., v. 47, no. 16, Aug. 1969. p. 1661-1670.

WOOLLEY, J.C. and K.W. BLAZEY. Reflectivity Spectra of GaSb-InSb and GaAs-InAs Alloys. J. OF PHYS. AND CHEM. OF SOLIDS., v. 25, no. 7, July 1964. p. 713-716.

WOOLLEY, J.C. and A.G. BRIGGS. Thermal Conductivity of GaSb-InSb Alloys. J. OF PHYS. AND CHEM. OF SOLIDS, v. 25, no. 8, Aug. 1964. p. 895-899.

WOOLLEY, J.C. and J.A. EVANS. Temperature Variation of Optical Energy Gap for GaSb-InSb. PROC. OF THE PHYS. SOC., v. 78, pt. 3, Sept. 1961. p. 354-360.

WOOLLEY, J.C. and C.M. GILLETT. Electrical Properties of GaSb-InSb Alloys. J. OF PHYS. AND CHEM. OF SOLIDS, v. 17, no. 1/2, Dec. 1960. p. 34-43.

WOOLLEY, J.C. and D.G. LEES. Equilibrium Diagrams with InSb as One Component. J. OF LESS COMMON METALS, v. 1, 1959. p. 192-198.

WROBEL, J.S. and H. LEVINSTEIN. Photoconductivity in InSb-GaSb and InAs-GaAs Alloys. INFRARED PHYS., v. 7, no. 4, Dec. 1967. p. 201-210.

ZEINALOV. S.A. and M.I. ALIEV. Investigation of the Thermal Conductivity of InSb-GaSb Solid Solutions. PHYS. STATUS SOLIDI, v. 22, no. 1, July 1967. p. 153-156.

ZEINALOV. S.A. et al. Structure of the Conduction Band and the Mechanism of Electron Scattering in $In_{0.5}Ga_{0.5}Sb$. SOVIET PHYS.-SEMICONDUCTORS, v. 4, no. 2, Aug. 1970. p. 324-326.

ZWERDLING, S. et al. Oscillatory Magnetoabsorption in GaSb JA-1149. PHYS. AND CHEM. OF SOLIDS, v. 9, no. 3/4, Mar. 1959. p. 320-324. (A)

ZWERDLING, S. et al. Oscillatory Magnetoabsorption in Semiconductors. PHYS. REV., v. 108, no. 6, Dec. 1967. p. 1402-1408. (B)

GALLIUM-INDIUM-ARSENIC SYSTEM

PROPERTY	SYMBOL	VALUE				UNIT	NOTES	TEMP.(°K)	REFERENCES
Formula		$Ga_xIn_{1-x}As$							
Symmetry		cubic							

Lattice Parameters	a_o	x	Giesecke & Pfister	Wagner	Hockings et al.	Cooper		NOTES	TEMP.(°K)	REFERENCES
		0	6.0584	6.057	6.055			InAs	300	
		0.04		6.056						
		0.74		6.055						
		0.90		6.054						
		3.7		6.045						
		4.2		6.043						
		4.9		6.037						
		6.2		6.035						
		11		6.012						
		15			6.000					
		31		5.926						
		34			5.920					
		35		5.900						
		53			5.845					
		66			5.790					
		85		5.710						
		90.5		5.688						
		96		5.665						
		100			5.650	5.64191		GaAs		

Melting Point	M.P.	x	Liquidus	Solidus		NOTES		REFERENCES
		0	943.3			InAs		Van den Boomgaard & Schol
		0	941.7		°C	melt-grown from high purity metals		Van Hook & Lenker, Steininger, Wu & Pearson
		35	1083.3					
		50	1125.0					
		75	1200.0					
		2		945		melt-grown, single crystals and single phase polycrystals, by liquid encapsulation		Wagner, Stringfellow & Green Rakov & Ufimtsev
		4		948				
		11		960				
		32		980				
		38		990				
		84		1075				
		90		1120				
		97		1200				
		100		1238		GaAs		Richman

Thermal Conductivity	k		300°K	500°K	700°K	UNIT	NOTES		REFERENCES
		0	0.250	0.125	0.901	W/cm °K	InAs melt-grown, zone leveled, polycrystalline		Hockings et al., Abeles, Abrahams et al.
		25	0.080	0.051	0.040				
		35	0.065	0.042	0.031				
		55	0.066	0.045	0.032				
		70	0.066	0.045	0.032				
		100	0.50	0.250	0.125		GaAs		

Dielectric Constant Optical	ε_∞	x	ε_∞	$\lambda_m(\mu)$	NOTES	TEMP.(°K)	REFERENCES
		0	8		InAs reflectivity meas. at 5μ on single phase, polycrystalline wafers	300	Thomas & Woolley
		28	11				
		66	10				
		79	10.5				
		32	11.3	7.9	optical dielectric constant, measured at the reflectivity min., λ_m on polycr., melt-grown wafers		Hockings et al.
		35	10.85	19.2			
		36	9.42	27.15			
		67	10.03	23.0			
		68	10.72	15.05			
		100	10.6	13.01	GaAs		

PROPERTY	SYMBOL	VALUE		UNIT	NOTES	TEMP.(°K)	REFERENCES

Electrical Resistivity — ρ

$\rho(10^{-3}$ ohm-cm)

x	77°K	300°K	500°K	800°K
20	1.4	1.8	1.9	1.5
37				2.5
60	1.2	1.45	1.75	15
78		2.5-9		
79		2.5-9		
82		2.5-9		
83				30
90		2.5-9		

Notes: polycrystalline or films, 1μ thick

References: Rosi et al., Coderre & Woolley, Jackson & Yee

Mobility — Electron μ_n **Hole** μ_p

x — $\mu_n(10^3 cm^2/V$ sec) carrier conc.(77,300°K)

x	77°K	300°K	$n(10^{17}cm^{-3})$
0	24.1	20.0	1.5
0.2	26.3	19.6	1.7
4	14.0	13.8	2.9
4.3	19.8	17.1	2.6
5	17.4	15.1	4.0
8.6	12.3	11.1	1.0
97	2.8	3.2	1.0
98	2.9	3.4	1.9
98.5	2.5	3.2	3.7
99	2.0	3.0	3.1
100	3.0	4.4	

Notes: InAs single crystals ... GaAs

References: Burdukov et al., (A, B, C), Dzhakhutashvili et al.

x	77°K	300°K	$n(10^{17}cm^{-3})$
0	48.3	25.7	.21
1	19	17.4	.10
3.4	24.6	16.9	.16
4	35	20	.24
11	29.4	17.8	.36
32	17.5	12.3	.62
38.5	-	14.9	-
96	3.2	3.1	12

Notes: single and polycrystals

References: Wagner

x	300°K
0	23
15	12
20	10
30	7
50-60	3
75	4
85	5
100	7

Notes: single crystal epitaxial films $n_n = 10^{15}$-10^{16}

References: Conrad et al. (A)

x	μ_n 300°K	μ_p 300°K	$n(cm^{-3})$
22-27		90	10^{19}
30-33	7000		10^{17}-10^{18}
35-36		70-80	3×10^{19}
58-64		50	7×10^{19}
66		80	10^{19}
68	3300		4×10^{18}
69	4170		10^{18}
70		70	10^{19}

Notes: melt-grown, polycr.

TEMP.: 300

References: Hockings et al.

Temperature Coeff. — μ_n

x	μ_n
20	$T^{0.5}$(80-200°K)
	T^{-1} (300-500°K)
	$T^{-1.5}$(750°K)

Notes: polycrystalline

TEMP.: 80-200

References: Babaev et al.

PROPERTY	SYMBOL	VALUE	UNIT	NOTES	TEMP.(°K)	REFERENCES

Effective Mass Electron — symbol m_n

x	m_n	UNIT	NOTES	TEMP.(°K)	REFERENCES
0	0.025	m_0	polycrystalline, single	300	Thomas & Woolley,
8	0.028		n-type, $n_n = 10^{18}$-10^{19}		Hockings et al.
18	0.031		reflectivity meas.		
28	0.034				
40	0.038				
47	0.043				
59	0.048				
66	0.049				
79	0.060				
90	0.065				
100	0.071				

Energy Gap
Direct (Γ_{15}-Γ_1) — E_0
Temp. Coeff. — dE_0/dT

x	E_0(eV)	dE_0/dT (-10^{-4}eV/°K)	NOTES	TEMP.(°K)	REFERENCES
0	0.4105		optical meas., single	0 (Extrap)	Adachi
0	0.40	4.65	InAs		Coderre & Woolley
10	0.45	4.95	n-type, polycrystalline,		
20	0.52	5.05	electrical meas.		
29	0.60	5.10	optical absorption		
38	0.67	5.25			
49	0.80	5.45			
60	0.90	5.70			
66	0.94	5.75			
83	1.15	6.0			
100	1.52	6.15	GaAs		
100	1.522		optical meas. on single crystals	0 (Extrap)	Sturge

x	E_0 100°K	210°K	300°K	UNIT	NOTES	TEMP.(°K)	REFERENCES
0	0.40	0.36	0.34	eV	photoconductivity meas. on		Taylor & Fortin,
12	0.50	0.43	–		polycrystalline wafers,		Wu & Pearson
24	0.58	0.56	–		$n = 5\times10^{16}$-2.5×10^{18}		
29	–	0.60	0.56				
39	0.73	0.67	0.64				
50	0.80	–	0.72				
56	0.90	0.82	–				
69	1.02	0.97	0.93				
75	1.1	1.05					
87	1.23	1.19					
100	1.48	1.42	1.41				

Energy Band Structure
Direct — E_0
E_1 (Λ_3-Λ_1)
Spin-Orbit Splitting (Δ_0, Δ_1)

x	E_0	$E_0+\Delta_0$	E_1	$E_1+\Delta_1$	NOTES	TEMP.(°K)	REFERENCES
0	0.350	0.796	2.524	2.791 eV	epitaxial single crystals	300	Williams & Rehn (A),
20	0.509	0.908	2.536	2.810	on (100) GaAs, 2-60µ thick		Thompson & Woolley
44	0.677	1.022	2.562	2.822	electroreflectance meas.		
60	0.821		2.61	2.84	$n = 10^{17}$-10^{18}		
75			2.672	2.945			
80	1.128	1.470	2.747	2.991			
83	1.167	1.50	2.75	3.00			
84	1.184		2.758	3.007			
100	1.416	1.755	2.904	3.132			

x	E_0'	$E_0'+\Delta_0'$	E_2	$E_2+\delta$	NOTES	TEMP.(°K)	REFERENCES
0	4.43		4.74	5.27	electroreflectance meas.	300	Thompson & Woolley
4	4.44		4.76	5.28	on single phase poly-		
14	4.44		–	–	crystals, $n_n = 7\times10^{16}$-10^{19}		
35	–		4.83	5.30			
44	–		4.84	5.30			
60	4.45	4.73	4.85	5.32			
79	4.46	4.73	4.91				
90	–	4.73	4.96				
100	4.47	4.67	4.99	5.34			

PROPERTY	SYMBOL		VALUE		UNIT	NOTES	TEMP.($\degree$K)	REFERENCES

Energy Band Structure Direct — E_o

x	E_o 300$\degree$K	E_o 77$\degree$K			
0	0.330	0.395	eV	InAs	Burdukov et al. (A, B, C)
0.2	0.343			optical absorption meas.	
4	0.361	0.430		on single crystals,	
4.3	0.363			$n_n = 1$-4×10^{17}	
5	0.370	0.450		also electroluminescence	
8.6	0.375	0.460		and photoluminescence meas.	
97	1.303	1.465			
98	1.308				
98.5	1.315	1.473			
99	1.328	1.488			
100	1.330	1.507		GaAs	

Phonon Branch Spectra
Transverse Optic TO_1, TO_2
Longitudinal Optic LO_1, LO_2

x	LO_1	LO_2	TO_1	TO_2			
0	–	29.5	–	26.6	meV	single crystal, epitaxial films, 40-100μ thick	300	Brodsky & Lucovsky
38	32.7	29.0	31.0	27.5			
50	33.8	28.9	31.7	28.2			
60	34.2	29.0	31.8	28.1			
61	–	–	–	–			
70	34.9	28.7	32.7	28.5			
78	35.6	–	32.8	28.5			
100	36.1	–	33.2	–			

Photoemission Work Function — ϕ
Quantum Efficiency — η

x	ϕ (eV)	Threshold (eV)	Quantum(η) Efficiency	Response	NOTES	TEMP	REFERENCES
75	0.8	1.05	0.004	130μ A/lm	polycrystalline, Cs_2O coated	300	Klein
70	0.7	0.95	0.002		polycrystalline, $n_n = 10^{19}$, Cs_2O-coated	300	Williams, Fisher et al.
86		1.18	0.05				
89		1.16	0.006	900	polycrystalline, 2.4×10^{18}	300	Uebbing & Bell
95		1.26	0.03	443	p-type, polycrystalline, CsF covered 8×10^{18}	300	Garbe
90		1.25	0.01		single crystal, Cs_2O, 1μ thick, vapor deposited, epitaxial film	300	Jackson & Yee
82		1.14	0.004				

Detectivity — D^*

x	D^* (cm Hz$^{0.5}$/W)	Peak Wavelength (μ)	NOTES	TEMP	REFERENCES
0	1.9×10^5	3.5	polycrystalline wafers, 5×10^{16}-2.5×10^{18}	300	Taylor & Fortin
39	2.8×10^5	1.87			
50	7.4×10^5	1.6			
69	2.1×10^5	1.3			
75	8.4×10^6	1.2			
100	2.4×10^6	0.87			

Laser Properties

	η%	$\lambda(\mu)$	Threshold Intensity	NOTES	TEMP	REFERENCES
80	0.6	1.09	2×10^5W/cm^2	vapor epitaxial film, 12-50μ thick $n_n = 5 \times 10^{17}$ Nd: YAG laser	300	Rossi et al.

Refractive Index — n

					REFERENCES
92-98	3.6				Basov et al.

GALLIUM-INDIUM-ARSENIC BIBLIOGRAPHY

ABELES, B. Lattice Thermal Conductivity of Disordered Semiconductor Alloys at High Temperatures. PHYS. REV., v. 131, no. 5, Sept. 1, 1963. p. 1906-1911.

ABRAHAMS, M.S. et al. Thermal, Electrical and Optical Properties of (In,Ga)As Alloys. J. OF PHYS. AND CHEM. OF SOLIDS, v. 10, no. 2/3, July 1959. p. 204-210.

ADACHI, E. Energy Band Parameters of Indium Arsenide at Various Temperatures. PHYS. SOC. OF JAPAN, J., v. 24, no. 5, May 1968. p. 1178.

BABAEV, R.M. et al. Electrical Properties of $In_{1-x}Ga_x$As Solid Solutions. SOVIET PHYS.-SEMICONDUCTORS, v. 2, no. 8, Feb. 1969. p. 1022-1023.

BASOV, N.G. et al. Properties of Injection Lasers at 0.8-1.1 Microns. SOVIET PHYS. TECH. PHYS., v. 12, no. 2, Aug. 1967. p. 250-257.

BRODSKY, M.H. and G. LUCOVSKY. Infrared Reflection Spectra of $Ga_{1-x}In_x$As: A New Type of Mixed-Crystal Behavior. PHYS. REV. LETTERS, v. 21, no. 14, Sept. 30, 1968. p. 990-993.

BURDUKOV, Yu.M. et al. Optical and Luminescence Properties of Single Crystals of In_xGa_{1-x}As Solid Solutions. SOVIET PHYS.-SEMICONDUCTORS, v. 4, no. 1, July, 1970. p. 138-141. (A)

BURDUKOV, Yu.M. et al. Photoluminescence and Electroluminescence of In_xGa_{1-x}As Solid Solutions. SOVIET PHYS.-SEMICONDUCTORS, v. 4, no. 10, Apr. 1971. p. 1633-1636. (B)

BURDUKOV, Yu.M. et al. Electrical Properties of Gallium-Rich Ga_xIn_{1-x}As Solid Solutions. SOVIET PHYS.-SEMICONDUCTORS, v. 4, no. 10, Apr. 1971. p. 1697-1701. (C)

CODERRE, W.M. and J.C. WOOLLEY. Conduction Bands of Ga_xIn_{1-x}As and $InAs_xSb_{1-x}$Alloys. CANADIAN J. OF PHYS., v. 48, no. 4, Feb. 1970. p. 463-469.

CONRAD, R.W. et al. Preparation of Epitaxial Ga_xIn_{1-x}As. ELECTROCHEM. SOC., J., v. 114, no. 2, Feb. 1967. p. 164-166. (A)

CONRAD, R.W. et al. Reflectivity Studies of Epitaxial Ga_xIn_{1-x}As. ELECTROCHEM. SOC., J., v. 113, no. 3, Mar. 1966. p. 287-289. (B)

COOPER, A.S. Precise Lattice Constants of Germanium, Aluminum, Gallium Arsenide, Uranium, Sulphur, Quartz and Sapphire. ACTA CRYST., v. 15, 1962. p. 578-582.

DZHAKHUTASHVILI, T.V. et al. Preparation and Physical Properties of Single Crystals of GaAs-InAs Solid Solutions. SOVIET PHYS.-SEMICONDUCTORS, v. 5, no. 2, Aug. 1971. p. 190-193.

FISHER, D.G. et al. Long-Wavelength Photoemission from $Ga_{1-x}In_x$As Alloys. APPLIED PHYS. LETTERS, v. 18, no. 9, May 1, 1971. p. 371-373.

GARBE, S. CsF,Cs as a Low Work Function Layer on the GaAs Photocathode. PHYS. STATUS SOLIDI, v. 2, no. 3, July 1970. p. 497-501.

GIESECKE, G. and H. PFISTER. Precision Determination of the Lattice Constants of III-V Compounds. ACTA CRYSTALLOGRAPHICA, v. 11, 1958. p. 369-371.

HOCKINGS, E.F. et al. Thermal and Electrical Transport in InAs-GaAs Alloys. J. OF APPLIED PHYS., v. 37, no. 7, June 1966. p. 2879-2887.

JACKSON, D.A. and E.M. YEE. Photoemission Yield Dependency on Bandgap Energy for GaInAs Alloys. IEEE PROC., v. 59, no. 1, Jan. 1971. p. 90-91.

JONES, D. and A.H. LETTINGTON. Pseudopotential Calculations of the Band Structure of GaAs, InAs and (Ga In)As Alloys. SOLID STATE COMM., v. 7, no. 18, Sept. 1969. p. 1319-1322.

KLEIN, W. Photoemission from Cesium Oxide Covered GaInAs. J. OF APPLIED PHYS., v. 40, no. 11, Oct. 1969. p. 4384-4389.

RAKOV, V.V. and V.B. UFIMTSEV. Phase Equilibrium in the Gallium Arsenide-Indium Arsenide System. RUSSIAN J. OF PHYS. CHEM., v. 43, no. 2, Feb. 1969. p. 267-268.

RICHMAN, D. Dissociation Pressures of Gallium Arsenide, Gallium Phosphide and Indium Phosphide and the Nature of III-V Melts. J. OF PHYS. AND CHEM. OF SOLIDS, v. 24, no. 9, Sept. 1963. p. 1131-1139.

ROSI, F.D. et al. Semiconducting Materials for Thermoelectric Power Generation. R.C.A. REV., v. 22, no. 1, Mar. 1961. p. 82-121.

ROSSI, J. et al. Optically Pumped Room Temperature In_xGa_{1-x}As Lasers. (M.I.T., LEXINGTON LINCOLN LAB., Solid State Res., 1971) Edited by: McWHORTER, A.L. p. 23-25. AD 736 501.

STEINIGER, J. Thermodynamics and Calculation of the Liquidus-Solidus Gap in Homogeneous, Monotonic Alloy Systems. J. OF APPLIED PHYS., v. 41, no. 6, May 1970. p. 2713-2724.

STRINGFELLOW, G.B. Calculation of Ternary Phase Diagrams of III-V Systems. J. OF PHYS. AND CHEM. OF SOLIDS, v. 33, no. 3, March 1972. p. 665-677.

STRINGFELLOW, G.B. and P.E. GREENE. Calculation of III-V Ternary Phase Diagrams: In-Ga-As and In-As-Sb. J. OF PHYS. AND CHEM. OF SOLIDS, v. 30, no. 30, July 1969. p. 1779-1791.

STURGE, M.D. Optical Absorption of Gallium Arsenide between 0.6 and 2.75 eV. PHYS. REV., v. 127, no. 3, Aug. 1962. p. 768-733.

TANSLEY, T.L. Spectral Response of P-N Hetero-Junctions. PHYSICA STATUS SOLIDI, v. 23, no. 1, 1967. p. 241-252. (A)

TANSLEY, T.L. Forward Current Injection Modulation of Photocurrent in P-N Hetero-Junctions. PHYS. STATUS SOLIDI, v. 24, no. 2, Dec. 1967. p. 615-622.

TAYLOR, A.E. and E. FORTIN. Photoconductivity in some III-V Alloys. CANADIAN J. OF PHYS., v. 48, no. 16, Aug. 1970. p. 1874-1878.

THOMAS, M.B. and J.C. WOOLLEY. Plasma Edge Reflectance Measurements in $Ga_xIn_{(1-x)}As$ and $InAs_xSb_{(1-x)}$ Alloys. CANADIAN J. OF PHYS., v. 49, no. 5, Aug. 1971. p. 2052-2060.

THOMPSON, A.G. and J.C. WOOLLEY. Electroreflectance Measurements on $Ga_xIn_{1-x}As$ Alloys. CANADIAN J. OF PHYS., v. 45, no. 8, Aug. 1967. p. 2597-2607.

UEBBING, J.J. and R.L. BELL. Improved Photoemitters Using GaAs and InGaAs. IEEE. PROC., v. 56, no. 9, Sept. 1968. p. 6124-1625.

VAN DEN BOOMGAARD, J. and K. SCHOL. The P-T-x Phase Diagrams of the Indium-Arsenic, Gallium-Arsenic and Indium-Phosphorus Systems. PHILIPS RES. REPTS., v. 12, no. 1, Apr. 1957. p. 127-140.

VAN HOOK, H.J. and E.S. LENKER. The System InAs-GaAs. AIME METALL., TRANS., SOC., v. 227, no. 1, Feb. 1963. p. 220-222.

WAGNER, J.W. Preparation and Properties of Bulk $In_{1-x}Ga_xAs$ Alloys. ELECTROCHEM. SOC., J., v. 117, no. 9, Sept. 1970. p. 1193-1196.

WILLIAMS, B.F. InGaAs-CsO, A Low Work Function (Less than 1.0 eV) Photoemitter. APPLIED PHYS. LETTERS, v. 14, no. 9, May 1, 1969. p. 273-275.

WILLIAMS, E.W. and V. REHN. Electroreflectance Studies of InAs, GaAs, and (Ga,In) As Alloys. PHYS. REV., v. 172, no. 3, Aug. 15, 1968. p. 798-810. (A)

WILLIAMS, E.W. and V. REHN. An Electroreflectance Interpretation of the $\Lambda_3\Lambda_1$ Transitions in InAs, GaAs, and (Ga,In) As Alloys. SOLID STATE COMM., v. 7, no. 7, Apr. 1969. p. 545-548. (B)

WOOLLEY, J.C. and K.W. BLAZEY. Reflectivity Spectra of GaSb-InSb and GaAs-InAs Alloys. J. OF PHYS. AND CHEM. OF SOLIDS, v. 25, no. 7, July 1964. p. 713-716.

WOOLLEY, J.C. et al. Electrical and Optical Properties of GaAs-InAs Alloys. PROC. OF THE PHYS. SOC., v. 77, Pt. 3, no. 495, Mar. 1, 1961. p. 700-704. (A)

WOOLLEY, J.C. et al. Optical Energy Gap Variation in $Ga_xIn_{1-x}As$ Alloys. CANADIAN J. OF PHYS., v. 46, no. 2, Jan. 1968. p. 157-159. (B)

WROBEL, J.S. and H. LEVINSTEIN. Photoconductivity in InSb-GaSb and InAs-GaAs Alloys. INFRARED PHYS., v. 7, no. 4, Dec. 1967. p. 201-210.

WU, T.Y. and G.L. PEARSON. Phase Diagram, Crystal Growth, and Band Structure of $In_xGa_{1-x}As$. J. OF PHYS. AND CHEM. OF SOLIDS, v. 33, no. 2, Feb. 1972. p. 409-415.

GALLIUM-INDIUM-PHOSPHORUS SYSTEM

PROPERTY	SYMBOL	VALUE			UNIT	NOTES	TEMP.($^\circ$K)	REFERENCES
Formula		$Ga_xIn_{1-x}P$						
Symmetry		cubic						
Lattice Parameter	a_o	x	a_o					
		0	5.86875		$\overset{\circ}{A}$	InP	300	Giesecke & Pfister
		17	5.82			polycrystalline ingots		Onton et al.
		23	5.78					
		36	5.72					
		39	5.71					
		50	5.66					
		59	5.62					
		83	5.53					
		86	5.51					
		100	5.4495			GaP		Pierron et al.
		0-50	$a_o = 5.8687-0.4182x +(0.0802x - 0.1614x^2)$ $\overset{\circ}{A}$					Onton et al.
		50-100	$a_o = 5.8687-0.418x$					
		44	5.68			pressed powder pellet of GaP and InP heated to homogenization		Foster & Scardefield
		46	5.67					
		68	5.57					
		72	5.55					
		86	5.50					
		90	5.49					
Melting Point	M.P.		M.P.					
		0	1070		$^\circ$C	InP		Koester & Ulrich
			Liquidus	Solidus				
		10	1170		$^\circ$C	solidus, thermal meas.($\pm$10°C)		Foster & Scardefield
		20	1240					
		30	1290					
		44	1330	1120		liquidus, calc.		Kajiyama
		46	1340	1125				
		50	1350					Hakki
		68	1415	1160				
		72	1425	1180				
		86	1450	1250				
		90	1460	1295				
		91		1305				
		95		1380				
		100	1465			GaP		Richman
Elastic Constants	s_{11}		s_{11}	$-s_{12}$				
	$-s_{12}$	0	1.307	0.400	10^{-12}cm^2/dyne	InP	300	Hakki et al.
		50	1.130	0.347		liquid phase		
		54	1.115	0.343		epitaxy deposition		
		63	1.090	0.333		on (111) GaAs.		
		66	1.080	0.330		n-type, p-n junction		
		74	1.047	0.322		diodes		
Mobility Electron	μ_n	x	μ_n(cm^2/V sec)	n_n(10^{16}cm^{-3})				
		57	600	50		single crystals,	300	Nuese et al.
		65	500	5		epitaxial vapor-		
		75	92	10		phase on (100) GaAs		

GALLIUM-INDIUM-PHOSPHORUS SYSTEM

PROPERTY	SYMBOL		VALUE			UNIT	NOTES	TEMP.(°K)	REFERENCES
Energy Gap Direct	E_{g_d}	x	E_{g_d}	E_{g_i}	(eV)				
		0	1.3511				InP optical measurement	300	Turner et al., Lorenz et al.
Indirect	E_{g_i}	0		1.2311			InP optical absorption	300	Onton et al. (B)
				$n_n(10^{16}\text{cm}^{-3})$					
		10	1.40	0.2			single crystals, epitaxial vapor phase on (100) GaAs, optical absorption, Schottky barrier and electro-luminescence meas.	300	Nuese et al.
		20	1.54						
		25	1.58						
		32	1.65						
		70	2.20	2.20					
		74		2.21					
		84		2.22					
		89		2.23					
		63		2.17			pressure meas. on p-n junction diode	300	Hakki et al.
		65		2.20			thermoreflectance meas. on epitaxial films, 10-50µ thick	300	Lettington et al.
		62		2.210			electroluminescence meas.	300	Laugier & Chevallier
		74		2.33	(cross-over)		photoluminescence meas. on single crystals	2	Onton & Chicotka
		74		2.26	(cross-over)			300	Onton et al.
		68		2.28			photoluminescence meas. on liquid phase epitaxy films	80	White et al.
		73	2.48	2.30					
		94-100		2.35					
		56		2.055			luminescence meas. on liquid phase epitaxy films, on GaAs, 4-7µ thick	300	Bachrach & Hakki
		58		2.080					
		64		2.174					
		100	2.78				GaP optical meas. on single crystal	290	Zallen & Paul, Abagyan & Subashiev
				2.259			optical meas. on perfect crystals	295	Dean et al.
Composition Coeff.	E_{g_d}		$1.340 + 0.668x + 0.758x^2$			eV	cathodoluminescence meas.	300	Onton et al. (A)
			$1.409 + 0.695x + 0.758x^2$				photoluminescence meas. on single crystals	2	Onton & Chicotka
	E_{g_i}		$2.321 + 0.17x$						

Energy Gap	E_{g_d}	x	5°K	200°K	270°K	297°K				
		60.5	2.215	2.164	2.133	2.105	eV	liquid phase epitaxy on GaAs, luminescence meas.		Bachrach & Hakki
	E_{g_i}		7.5°K	88°K	100°K	300°K				
		64	2.28	2.271	2.262	2.174				

Temperature Coeff.	dE_{g_d}/dT	60.5,64	-4.3×10^{-4} eV/°K					100-300	Bachrach & Hakki

PROPERTY	SYMBOL	VALUE	UNIT	NOTES	TEMP.(°K)	REFERENCES
Energy Gap Pressure Coeff.	dE_{g_d}/dP	$\frac{x}{50}$ 13×10^{-6} 54 12	eV/kg-cm^{-2}	liquid phase epitaxy deposition on (111) GaAs, n-type, p-n junction diodes	300	Hakki et al.
	dE_{g_i}/dP	63 -1 66 -1.2 74 -1.3				
Deformation Potential	D_d	50 9.8 54 9.3	eV		300	Hakki et al.
	D_i	63 0.80 66 0.95 74 1.05				

Energy Band Structure	x	E_o	$E_o+\Delta_o$	E_1	$E_1+\Delta_1$	$E_o{'}$	$E_o{'}+\Delta_o{'}$	E_2	$E_2+\delta$		
	57	2.042	2.145	3.306	3.52	4.78	5.04	5.20	5.65	electroreflectivity meas. on single crystals at 300°K	Alibert et al.

Phonon Spectra (X-point) Longitudinal Optic LO_1 Transverse Optic TO_1	x	LO_1	TO_1	UNIT	NOTES	TEMP.	REFERENCES
	0	43.4	38.1	meV	optical meas. on single and polycrystalline liquid growth samples	300	Lucovsky et al.
	20	45.2	39.0				
	42	48.3	41.0				
	65	48.7	42.4				
	82	49.3	44.0				
	100	49.6	45.0				

Refractive Index	n	$\frac{n}{64}$ 3.46		liquid phase epitaxy on GaAs	300	Bachrach & Hakki

Laser Properties	x	Wavelength (Å)	Threshold Current Density (A/cm^2)	Efficiency (%)	NOTES	TEMP.	REFERENCES
	27	7620	5.5×10^4	0.74	single crystals grown by slow solution, annealed at 925°C, cleaved into platelets 1-5μ thick, $n_n = 4.5 \times 10^{18}$, (Te-doped)	77	Scifres et al. (A), Burnham et al. (A, B), Macksey et al.
	59	5800	6.7×10^4				
	53	6600	1×10^4		Te-doped platelets	300	Scifres et al. (B)
	60	6000					
Electroluminescent Diodes	57	6105-6180	$4-6 \times 10^3$	2.5	vapor phase epitaxial deposition on (100) GaP, p-n junction 5-10μ deep, 310 fL/A cm^2	80	Nuese et al.
	68	5950	5.3		p-n junctions prepared by very slow cooling, Te- and Zn-doped. 720 ftL/A cm^2	293	Okuno et al.
	62	6000					

GALLIUM-INDIUM-PHOSPHIDE BIBLIOGRAPHY

ABAGYAN, S.A. and V.K. SUBASHIEV. Direct Transitions and Spin-Orbital Splitting of the Valence Band in Gallium Phosphide. SOVIET PHYS. SOLID STATE, v. 6, no. 10, Apr. 1965. p. 2529-2530.

ALIBERT, C. et al. Study of Gallium Indium Phosphides by Electroreflectivity (In Fr.). ACAD. DES SCI., C.R., v. 274, no. 9, Ser. B, Feb. 28, 1972. p. 653-655.

BACHRACH, R.Z. and B.W. HAKKI. Radiative Processes in Direct and Indirect Band Gap $In_{1-x}Ga_xP$. J. OF APPLIED PHYS., v. 42, no. 12, Nov. 1971. p. 5102-5108.

BURNHAM, R.D. et al. Stimulated Emission in $In_{1-x}Ga_xP$. APPLIED PHYS. LETTERS, v. 17, no. 10, Nov. 15, 1970. p. 430-432. [A]

BURNHAM, R.D. et al. Spectral Behavior, Carrier Lifetime, and Pulsed and cw Laser Operation (77°K) of $In_{1-x}Ga_xP$. APPLIED PHYS. LETTERS, v. 18, no. 4, Feb. 15, 1971. p. 160-162. [B]

DEAN, P.J. et al. Intrinsic Optical Absorption of Gallium Phosphide Between 2.33 and 3.12 eV. J. OF APPLIED PHYS., v. 38, no. 9, Aug. 1967. p. 3551-3556.

FOSTER, L.M. and J.E. SCARDEFIELD. The Solidus Boundary in the GaP-InP Pseudobinary System. ELECTROCHEM. SOC., J., v. 117, no. 4, Apr. 1970. p. 534-536.

GIESECKE, G. and H. PFISTER. Precision Determination of the Lattice Constant of III-V Compounds. ACTA CRYSTALLOGRAPHICA, v. 11, 1958. p. 369-371.

HAKKI, B.W. Growth of $In_{1-x}Ga_xP$ p-n Junction by Liquid phase Epitaxy. ELECTROCHEM. SOC., J., v. 118, no. 9, Sept. 1971. p. 1469-1473.

HAKKI, B.W. et al. Band Structure of InGaP from Pressure Experiments. J. OF APPLIED PHYS., v. 41, no. 13, Dec. 1970. p. 5291-5296.

ITOH, H. et al. Reproducible Preparation of Homogeneous $In_{1-x}Ga_xP$ Mixed Crystals. APPLIED PHYS. LETTERS, v. 19, no. 9, Nov. 1, 1971. p. 348-349.

KAJIYAMA, K. The In-Ga-P Ternary Phase Diagram. JAPAN. J. OF APPL. PHYS., v. 10, no. 5, May 1971. p. 561-565.

KOESTER, W. and W. ULRICH. Isomorphy in III-V Compounds (In Ger.). Z. FUER METALLKUNDE, v. 49, no. 7, July 1956. p. 365-367.

KYSER, D.S. et al. Indirect Transitions in GaP via Electroreflectance and Electroabsorption. AMERICAN PHYS. SOC., BULL., Ser. 2, v. 15, no. 3, Mar. 1970. p. 288.

LAUGIER, A. and J. CHEVALLIER. About the Band Structure of $Ga_xIn_{1-x}P$ Alloys. SOLID STATE COMMUNICATIONS, v. 10, no. 4, Feb. 1972. p. 353-356.

LETTINGTON, A.H. et al. Thermoreflectance Studies of Thin Epitaxially Deposited (InGa)P Alloys. J. OF PHYS., C, v. 4, no. 12, Aug. 1971. p. 1534-1539.

LOGAN, R.A. et al. Electroluminescence in $GaAs_xP_{1-x}$, $In_xGa_{1-x}P$, and $Al_xGa_{1-x}P$ Junctions with x about 0.01. J. OF APPLIED PHYS., v. 42, no. 6, May 1971. p. 2328-2335.

LORENZ, M.R. et al. Band Structure and Direct Transition Electroluminescence in the $In_{1-x}Ga_xP$ Alloys. APPLIED PHYS. LETTERS, v. 13, no. 12, Dec. 15, 1968. p. 421-423.

LUCOVSKY, G. et al. Long-Wavelength Optical Phonons in $Ga_{1-x}In_xP$. PHYS. REV., B, Ser. 3, v. 4, no. 6, Sept. 15, 1971. p. 1945-1949.

MABBITT, A.W. The Cathodoluminescence of $Ga_xIn_{1-x}P$ Alloys. SOLID STATE COMMUNICATIONS, v. 9, no. 3, Feb. 1971. p. 245-247.

MACKSEY, H.M. et al. $In_{1-x}Ga_xP$ p-n Junction Lasers. APPLIED PHYS. LETTERS, v. 19, no. 8, Oct. 15, 1971. p. 271-273.

NUESE, C.J. et al. The Preparation and Properties of Vapor-Grown $In_{1-x}Ga_xP$. METALLURGICAL TRANS., v. 2, no. 3, Mar. 1971. p. 789-794.

ONTON, A. and R.J. CHICOTKA. Photoluminescence Processes in $In_{1-x}Ga_xP$ at 2°K. PHYS. REV. B, Ser. 3, v. 4, no. 6, Sept. 15, 1971. p. 1847-1853.

ONTON, A. et al. Electronic Structure and Luminescence Processes in $In_{1-x}Ga_xP$ Alloys. J. OF APPLIED PHYS., v. 42, no. 9, Aug. 1971. p. 3420-2432. [A]

ONTON, A. et al. Direct Optical Observation of the Subsidiary X_{1c} Conduction Band and Its Donor Levels in InP. AMERICAN PHYS. SOC., BULL., v. 17, no. 3, Ser. 11, Mar. 1972. p. 326. [B]

PANISH, M.B. and H.C. CASEY, JR. Temperature Dependence of the Energy Gap in Gallium Arsenide and Gallium Phosphide. J. OF APPLIED PHYS., v. 40, no. 1, Jan. 1969. p. 163-167.

PIERRON, E.D. et al. Coefficient of Expansion of Gallium Arsenide, Gallium Phosphide and Gallium Arsenic Phosphide Compounds from 62 to 200°C. J. OF APPLIED PHYS., v. 38, no. 12, Nov. 1967. p. 4669-4671.

RICHMAN, D. Dissociation Pressure of Gallium Arsenide, Gallium Phosphide and Indium Phosphide and the Nature of III-V Melts. J. OF PHYS. AND CHEM. OF SOLIDS, v. 24, no. 9, Sept. 1963. p. 1131-1139.

RODOT, H. et al. Preparation and Optical Properties of Gallium-Indium Phosphorus Alloys (In Fr.). ACAD. DES SCI., C.R., v. 269, no. 9, Ser. B, Sept. 1, 1969. p. 381-384.

SCIFRES, D.R. et al. Optically Pumped $In_{1-x}Ga_xP$ Platelet Lasers from the Infrared to the Yellow (8900-5800 $\overset{\circ}{A}$, 77°K). J. OF APPLIED PHYS., v. 43, no. 3, Mar. 1972. p. 1019-1022. [A]

SCIFRES, D.R. et al. Optically Pumped Volume-Excited cw Room Temperature $In_{1-x}Ga_xP$ (x <0.60) Platelet Lasers. APPLIED PHYS. LETTERS, v. 20, no. 5, Mar. 1972. p. 184-186. [B]

TURNER, W.J. et al. Exciton Absorption and Emission in Indium Phosphide. PHYS. REV., v. 136, no. 5A, Nov. 1964. p. A1467-A1470.

WHITE, A.M. et al. Applications of Photoluminescence Excitation Spectroscopy to the Study of Indium Gallium Phosphide Alloys. J. OF PHYS., D, v. 3, no. 9, Sept. 1970. p. 1322-1328.

WHITE, A.M. et al. Infra-Red and Visible Photoluminescence in $In_{1-x}Ga_xP$. PHYS. STATUS SOLIDI, v. 30, no. 2, Dec. 1968. p. K125-K126.

WILLIAMS, E.W. et al. Evidence for a Pair Band Associated with Zinc in (In,Ga)P. SOLID STATE COMMUNICATIONS, v. 8, no. 7, Apr. 1970. p. 501-503.

WILLIAMS, E.W. et al. (In,Ga)P Alloys: Photoluminescence Excitation and Cathodoluminescence of Zinc Doped Indirect Gap Alloys. J. OF PHYS., C, v. 3, no. 2, Feb. 1970. p. L55-L57.

ZALLEN, R. and W. PAUL. Band Structure of Gallium Phosphide from Optical Experiments at High Pressures. PHYS. REV., v. 134, no. 6A, June 1964. p. A1628-A1641.

NUESE, C.J. et al. Orange Laser Emission and Bright Electroluminescence from $In_{1-x}Ga_xP$ Vapor-Grown p-n Junctions. APPLIED PHYS. LETTERS, v. 20, no. 11, June 1972. p. 431-434.

OKUNO, Y. et al. Bright Yellow Luminescence from $In_{1-x}Ga_xP$ p-n Junctions. JAPAN. J. OF APPL. PHYSICS, v. 11, no. 5, May 1972. p. 757.

INDIUM-ARSENIC-ANTIMONY SYSTEM

PROPERTY	SYMBOL	VALUE	UNIT	NOTES	TEMP.(°K)	REFERENCES
Formula		$InAs_xSb_{1-x}$				
Symmetry		cubic				

Lattice Parameters a_o

x	a_o	UNIT	NOTES	REFERENCES
0	6.47877	$\overset{\circ}{A}$	InSb	Giesecke & Pfister
0	6.48		melt-grown and zone refined	Woolley & Warner (A)
10	6.45			
20	6.41			
35	6.36			
40	6.34			
50	6.30			
60	6.25			
65	6.21			
70	6.18			
75	6.15			
80	6.13			
85	6.11			
95	6.07			
100	6.06		InAs	
100	6.0584			Giesecke & Pfister

Melting Point M.P.

x	Liquidus	Solidus	UNIT	NOTES	REFERENCES
0	525.2		°C	InSb	Bednar & Smirous
15	708	535			Shih & Peretti, Woolley & Smith, Steininger
45	838	551			
69	889	570			
80	914	600			
90	930	730			
95	936	820			
100	943			InAs	Van den Boomgaard & Schol

Dielectric Constant — Optical ϵ_∞

x	ϵ_∞	NOTES	TEMP.(°K)	REFERENCES
5	12.8	homogeneous, n-type, polycrystalline, reflectivity meas. at 6μ	300	Thomas & Woolley
22	12.0			
30	11.5			
84	9.2			

Electrical Resistivity ρ

x	200°K	300°K	600°K	UNIT	NOTES	REFERENCES
37	5	2	0.5	10^{-3} Ω-cm	polycrystalline, zone refined	Coderre & Woolley (B)
53	2	2	0.4			
80	4	4	1			

Mobility — Electron μ_n

x	300°K μ_n ($10^3 cm^2/Vsec$)	300°K n_n ($10^{16} cm^{-3}$)	77°K μ_n	77°K n_n ($10^{16} cm^{-3}$)	NOTES	TEMP.(°K)	REFERENCES
98	25	3.2	33	3.0	single crystals		Basov et al.
	12	0.016	15	0.016			
0	57.5	13			InSb macrocrystalline, zone refined	300	Kudman & Ekstrom
9	47.4	6.5					
31.5	79.6	4.2					
52	5.9	3.6					
64	19.8	3.4					
82	28.4	3.7					

INDIUM-ARSENIC-ANTIMONY SYSTEM

PROPERTY	SYMBOL	VALUE			UNIT	NOTES	TEMP.(°K)	REFERENCES

Mobility Electron — μ_n

x	$\mu_n(10^4 cm^2/V sec)$			NOTES	REFERENCES
	273°K	473°K	773°K		
0	4.5	3.0	0.8	InSb	Coderre & Woolley (A)
28	4.0	2.5	0.7	n-type, polycrystalline	
38	3.2	1.7	0.5		
54	3.2	2.1	0.5		
68	2.2	1.6	0.6		
79	2.0	1.2	0.6		
100	1.3	1.0	0.6	InAs	

Effective Mass Electron m_n, **Heavy Hole** m_{hp} (UNIT: m_0)

x	m_n	m_{hp}	$n(10^{16} cm^{-3})$	NOTES	TEMP.(°K)	REFERENCES
9	0.0115	0.49	6.5	macrocrystalline, electrical meas.	300	Kudman & Ekstrom
31.5	0.0127	0.50	4.2			
52	0.0129	0.50	3.6			
64	0.0129	0.46	3.4			
82	0.0153	0.45	3.7			

Electron m_n

x	250°K	300°K	333°K	400°K
31.5	0.0127	0.0127	0.0155	0.0175

x	m_n	$n(10^{18} cm^{-3})$	NOTES	TEMP.(°K)	REFERENCES
10	0.031	3.4	vapor deposited films, reflectivity meas.	300	Potter & Kretschmar
10	0.051	10			
25	0.027	1.8			
55	0.038	4.0			
70	0.044	5.5			
100	0.066	8.5	InAs		
0	0.0133	6.3	InSb	300	Thomas & Woolley, Aubin & Woolley, Van Tongerloo & Woolley
5	0.0109	4.0	homogeneous, n-type, reflectivity, Faraday rotation or electrical meas.		
15	0.0107	4.2			
22	0.0102	5.0			
30	0.0105	6.0			
44	0.0094	0.08			
53.5	0.0112	0.9			
63.5	0.0128	0.1			
70	0.0120	7.2			
76	0.0158	8.0			
80	0.0150	8.4			
84	0.0182	7.5			
89	0.0181	5.3			
94	0.0260	3.6			
100	0.0250	12.0	InAs		

Energy Gap — E_g

x	E_g	UNIT	NOTES	TEMP.(°K)	REFERENCES
0	0.2355	eV	InSb magnetoabsorption at 4°K, pure single crystals	0	Pidgeon & Brown
0	0.180		optical and magnetoabsorption meas. on single crystals	300	Lukes & Schmidt, Zwerdling et al.
9	0.195		electrical meas. on n-type macrocrystalline samples	0	Kudman & Ekstrom
31.5	0.218				
52	0.226				
64	0.228				
82	0.28				
10	0.25		vapor deposited films, reflectivity meas.	300	Potter & Kretschmar
25	0.30				
55	0.33				
70	0.34				
100	0.36		InAs		

PROPERTY	SYMBOL	VALUE		UNIT	NOTES	TEMP.(°K)	REFERENCES

Energy Gap — E_g

	x	E_g 0°K*	E_g 300°K**				
	0	0.242	0.0175	eV	InSb		*Coderre & Woolley (A)
	9	0.210	0.140				
	15	0.201	0.135				**Woolley & Warner (B)
	25.5	0.177	0.110				
	38	0.165	0.105				
	53.5	0.175	-				
	79	0.277	0.224				
	88	0.361	0.26				
	95	0.418	0.325				
	100	0.448	0.35		InAs		
	100	0.4105	0.356		magnetoabsorption on single crystal		Adachi

E_g	x	E_g				TEMP.(°K)	
	35	0				256	Coderre & Woolley (B)

Energy Band Structure

	x	E_1	$E_1+\Delta_1$	E_0'	E_2	UNIT	NOTES	TEMP.(°K)	REFERENCES
	0	2.5	2.8	4.4	4.7	eV	macrocrystalline, electroreflectivity meas.	300	Vishnubhatla et al., Cardona et al.
	10	2.4	2.7	4.2	4.5				
	20	2.3	2.6	4.1	4.45				
	30	2.25	2.5	4.0	4.35				
	35	2.15	2.5	3.9	4.3				
	65	1.9	2.45	-	-				
	80-100	1.8	2.4	-	4.1				
	100			3.1					

Phonon Branch Spectra Transverse Optic — TO

	x	TO_1	TO_2	UNIT	NOTES	TEMP.(°K)	REFERENCES
	20	25.8	22.9	meV	polycrystalline wafers, far infrared reflectivity meas.	300	Lucovsky & Chen
	25	26.8	23.3				
	85	28.5					

Seebeck Coeff. — Q

	x	Q	UNIT	NOTES	TEMP.(°K)	REFERENCES
	5.5	287.6	μV/°K	n-type, polycrystalline	300	Aubin & Woolley
	14	248.4				
	27.5	218.5				
	40	183.2				
	48	199.0				
	58	201.0				
	84	214.9				
	88	177.0				
	96	264.3				

Laser Properties

	x	TCD	Wavelength	NOTES	TEMP.(°K)	REFERENCES
	>98	80-1000 A/cm^2	3.17μ	single crystal, p-n junction, 20-30μ deep, 5-8x10^{-3} cm^2 area	77	Basov et al.

INDIUM-ARSENIC-ANTIMONY BIBLIOGRAPHY

ADACHI, E. Energy Band Parameters of Indium Arsenide at Various Temperatures. PHYS. SOC. OF JAPAN, J., v. 24, no. 5, May 1968. p. 1178.

ADLER, P.N. Pressure-Induced Transformation Behavior of Some Quasi-Binary Alloys of InSb. J. OF PHYS. AND CHEM. OF SOLIDS, v. 30, no. 5, May. 1969. p. 1077-1082.

AUBIN, M.J. and J.C. WOOLLEY. Electron Scattering in $InAs_xSb_{1-x}$ Alloys. CANADIAN J. OF PHYS., v. 46, no. 10, Pt. 1, May 15, 1968. p. 1191-1198.

BASOV, N.G. et al. Semiconductor p-n Junction Lasers in the $InAs_{1-x}Sb_x$ System. SOVIET PHYS.-SOLID STATE, v. 8, no. 4, Oct. 1966. p. 847-849.

BEDNAR, J. and K. SMIROUS. The Melting Point of Gallium and Indium Antimonide (In Ger.). CZECH. J. PHYS., v. 5, no. 4, 1955. p. 546.

CARDONA, M. Optical Properties of Semiconductors above the Fundamental Absorption Edge. INTERNAT. CONF. ON SEMICONDUCTOR PHYS., PROC., 7th, Paris, 1964. Edited by: M. HULIN, N.Y. Academic Press. p. 181-196.

CARDONA, M. et al. Electroreflectance at a Semiconductor-Electrolyte Interface. PHYS. REV., v. 154, no. 3, Feb. 1967. p. 696-720.

CODERRE, W.M. and J.C. WOOLLEY. Electrical Properties of $InAs_xSb_{1-x}$ Alloys. CANADIAN J. OF PHYS., v. 46, no. 10, Pt. 1, May 15, 1968. p. 1207-1214. (A)

CODERRE, W.M. and J.C. WOOLLEY. Conduction Bands of $Ga_xIn_{1-x}As$ and $InAs_xSb_{1-x}$ Alloys. CANADIAN J. OF PHYS., v. 48, no. 4, Feb. 1970. p. 463-469. (B)

GIESECKE, G. and H. PFISTER. Precision Determination of the Lattice Constants of III-V Compounds. ACTA CRYSTALLOGRAPHICA, v. 11, 1958. p. 369-371.

KUDMAN, I. and L. EKSTROM. Semiconducting Properties of InSb-InAs Alloys. J. OF APPLIED PHYS., v. 39, no. 7, June 1968. p. 3385-3388.

LUCOVSKY, G. and M.F. CHEN. Longwave Optical Phonons in the Alloy Systems: $Ga_{1-x}In_xAs$, $GaAs_{1-x}Sb_x$ and $InAs_{1-x}Sb_x$. SOLID STATE COMM., v. 8, no. 17, Sept. 1970. p. 1397-1401.

LUKES, F. and E. SCHMIDT. The Fine Structure and the Temperature Dependence of the Reflectivity and Optical Constants of Germanium, Silicon and III-V Compounds. INTERNAT. CONF. ON THE PHYSICS OF SEMICONDUCTORS, PROC., Exeter, July 1962. Edited by: A.C. STICKLAND, London, Inst. of Phys. and the Phys. Soc. p. 389-394.

PIDGEON, C.R. and R.N. BROWN. Interband Magneto-Absorption and Faraday Rotation in Indium Antimonide. PHYS. REV., v. 146, no. 2, June 1966. p. 575-583.

POTTER, R.F. Optical Properties of Multisource Thermally Evaporated III-V Semiconductor Compounds. APPLIED OPTICS, vo. 5, no. 1, Jan. 1966. p. 35-40.

POTTER, R.F. and G.G. KRETSCHMAR. Optical Properties of $InAs_ySb_{1-y}$ Layers Prepared by Thermal Evaporation. INFRARED PHYS., v. 4, no. 1, Mar. 1964. p. 57-65.

SHIH, C. and E.A. PERETTI. The Phase Diagram of the System InAs-Sb. AMERICAN SOC. FOR METALS, TRANS., v. 46, 1954. p. 389-396.

STEININGER, J. Thermodynamics and Calculation of the Liquidus-Solidus Gap in Homogeneous, Monotonic Alloy Systems. J. OF APPLIED PHYS., v. 41, no. 6, May 1970. p. 2713-2724.

STRINGFELLOW, G.B. Calculation of Ternary Phase Diagrams of III-V Systems. J. OF PHYS. AND CHEM. OF SOLIDS, v. 33, no. 3, Mar. 1972. p. 665-677.

STRINGFELLOW, G.B. and P.E. GREENE. Calculation of III-V Ternary Phase Diagrams: In-Ga-As and In-As-Sb. J. OF PHYS. AND CHEM. OF SOLIDS, v. 30, no. 7, July 1969. p. 1779-1791.

THOMAS, M.B. and J.C. WOOLLEY. Plasma Edge Reflectance Measurements in $Ga_xIn_{(1-x)}As$ and $InAs_xSb_{(1-x)}$ Alloys. CANADIAN J. OF PHYS., v. 49, no. 5, Aug. 1971. p. 2052-2060.

VAN DEN BOOMGAARD, J. and K. SCHOL. The P-T-x Phase Diagrams of the Indium-Arsenic, Gallium-Arsenic and Indium-Phosphorus Systems. PHILIPS RES. REPTS., v. 12, no. 1, Apr. 1957. p. 127-140.

VAN TONGERLOO, E.H. and J.C. WOOLLEY. Free-Carrier Faraday Rotation in $InAs_xSb_{1-x}$ Alloys. CANADIAN J. OF PHYS., v. 46, no. 10, Pt. 1, May 15, 1968. p. 1199-1206.

VISHNUBHATLA, S.S. et al. Electroreflectance Measurements in Mixed III-V Alloys. CANADIAN J. OF PHYS., v. 47, no. 16, Aug. 1969. p. 1661-1670.

WOOLLEY, J.C. and B.A. SMITH. Solid Solution in III-V Compounds. PHYS. SOC., PROC., v. 72, 1958. p. 214-223.

WOOLLEY, J.C. and J. WARNER. Preparation of InAs-InSb Alloys. ELECTROCHEM. SOC., J., v. 111, no. 10, Oct. 1964. p. 1142-1145. (A)

WOOLLEY, J.C. and J. WARNER. Optical Energy-Gap Variation in AnAs-InSb Alloys. CANADIAN J. OF PHYS., v. 42, no. 10, Oct. 1964. p. 1879-1885. (B)

ZWERDLING, S. et al. Oscillatory Magnetoabsorption in Semiconductors. PHYS. REV., v. 108, no. 6, Dec. 1967.

PROPERTY	SYMBOL	VALUE		UNIT	NOTES	TEMP.(°K)	REFERENCES
Formula		$InAs_xP_{1-x}$					
Symmetry		cubic					

Lattice Parameter a_o

x	a_o	UNIT	NOTES	REFERENCES
0	5.86875	Å	InP	Giesecke & Pfister
0	5.8691		sealed-tube, iodine-	Thompson et al.
10	5.88925		transport, vapor-growth;	
20	5.9125		samples from first de-	
30	5.9303		posited portion of ingot	
40	5.9540			
50	5.9707			
60	5.9866			
70	6.0049			
80	6.02205			
90	6.0411			
100	6.0583		InAs	
100	6.0584			Giesecke & Pfister

Melting Point M.P.

x	M.P. Liquidus	M.P. Solidus	UNIT	NOTES	REFERENCES
0	1062		°C	InP 60 atm phosphorus pressure	Van den Boomgaard & Schol
0	1062			macrocrystalline	Willardson et al., Antypas & Yep
15.6		1037			
26.9	1035				
29.5		1020			
30.4		1016			
31.8		1009			
33	1021				
42	1016				
55.2	1009	993			
59.3	993				
90.6		951			
94.5	951				
100	943				
100	943			InAs	Van den Boomgaard & Schol

Thermal Conductivity

x	300°K	500°K	1000°K	UNIT	NOTES	REFERENCES
60	0.17	0.10	0.07	W/cm °K	melt-grown, poly-	Bowers et al.
80	0.15	0.09	0.06		crystalline	
90	0.12	0.083	0.055			
95	0.11	0.08	0.05			

Dielectric Constant
Optical ε_∞
Static ε_o

x	ε_∞	ε_o	NOTES	REFERENCES
60	11.1	14.6	calc.	Ehrenreich
80	11.3	14.4		

Electrical Resistivity ρ

$\rho(10^{-3}\Omega\text{-cm})$

x	300°K	625°K	770°K	1100°K	$n_n(10^{17}cm^{-3})$	NOTES	REFERENCES
10	3.2				10	single crystal	Willardson et al.A
60		10	10	2.0	2	polycrystalline	Bowers et al.
80		9	3	0.7	10		
90		5	1.6	0.6	60		
95		2.5	1.2	0.5	4		

PROPERTY	SYMBOL	VALUE			UNIT	NOTES	TEMP.($^{\circ}$K)	REFERENCES
			μ_n(cm^2/V sec)					
		x	77°K	300°K				
Mobility Electron	μ_n	0	19000	3000		single crystal, epitaxial film, vapor-deposited, 50-150μ thick $n_n = 5\times10^{15}$-10^{16}		Tietjen et al., Makhalov & Melik-Davtyan
		20		3500				
		40	15000	3500				
		50	25000	4000-7500				
		70	32000	7000				
		80	45000	12000				
		90	50000	13000				
		100	120000	32000		InAs		
		17	19000	7000		macrocrystalline		Willardson et al. (A, B)
		35	20000	8000				
		38	23000	8000				
		1000	75000	30000				
		0-25		2000-3000		liquid phase epitaxy layers, 20μ thick, $n_n = 4\times10^{16-17}$		Antypas & Yep
	μ_n	0-20	200			open tube, vapor epitaxial film on semi-insulating GaAs, (111) (110), (100) $n_n \sim$ 1-5$\times10^{16}$cm^{-3}		Allen & Mehal
		40-50	900					
		75	3000					
		90	6000-7000					
		95	8000					
		100	10000-11000					
			m_n	$n_n(10^{16}$cm$^{-3})$				
Effective Mass Electron	m_n	40	0.08	4.1		n-type, polycrystalline optical meas. at 1-35μ	300	Oswald
		60	0.06	1.0				
		80	0.045	0.7				
		10	0.105-0.11	34		Faraday rotation at 3-13μ on polycrystalline material	300	Makhalov & Melik-Davtyan
		20-30	0.10	2-9				
		40	0.090-0.095	30				
		50-60	0.065	2-4				
		70	0.055	2.0				
		80	0.05	1.6				
		85	0.04					
		90	0.03	0.97				
			E_g	$E_g + \Delta_o$				
Energy Gap Direct Spin-Orbit Splitting	E_g Δ_o	0	1.34		eV	InP electroreflectivity	300	Cardona et al.
				1.45		electroreflectivity	300	Shaklee et al., Irzekevicius et al.
		7	1.25	1.4		electroreflectivity, n-type, polycrystalline, 10^{16}-2.5$\times10^{17}$	300	Irzekevicius et al.
		16	1.15	1.3				
		43	0.80	1.5				
		57	0.74	1.2				
		66	0.68	0.95		optical meas.	300	Dubrovskii
		82	0.52	0.88				
		90	–	0.76		liquid phase epitaxy films, optical meas.	300	Antypas & Yep
		100	0.35	0.74				
		100	0.359	0.739		InAs magnetoabsorption	300	Pidgeon et al. (A, B)

PROPERTY	SYMBOL	VALUE	UNIT	NOTES	TEMP.(°K)	REFERENCES

Energy Band Structure (x)

x	E_1 77°K	E_1 300°K	$E_1+\Delta_1$ 77°K	$E_1+\Delta_1$ 300°K	E_o' 300°K	E_2	$E_2+\delta$ (eV)
0	3.25	3.18	3.38	3.30	4.7	5.02	5.6
10	3.18	3.10	3.35	3.28	4.65		
20	3.14	3.03	3.35	3.22		4.98	5.51
30	3.05	2.95	3.25	3.17	4.64		
40	2.95	2.85	3.18	3.10			
50	2.88	2.83	3.14	3.05			
60	2.82	2.75	3.06	3.0	4.56	4.82	5.40
70	2.77	2.65	3.02	2.92			
80	2.72	2.63	2.98	2.90			
90	2.67	2.58	2.95	2.85		4.75	5.27
100	2.62	2.52	2.90	2.83	4.42	4.72	5.26

NOTES: halogen, vapor-transport, closed-tube, polycr., electroreflectivity meas., $n_n = 10^{16}-10^{17}$ (100 = InAs)

REFERENCES: Thompson et al., Kavaliauskas & Sileika, Vishnubhatla et al.

Composition Coeff.

$$E_o = 0.36 + 0.82(1-x) + 0.16(1-x)^2 \ \text{eV}$$

$$E_o + \Delta_o = 0.78 + 0.46(1-x) + 0.2\,(1-x)^2$$

NOTES: electroreflectance meas. at 300°K on polycrystalline samples, $n_n = 10^{16}-2.5\times10^{17}\,\text{cm}^{-3}$

REFERENCES: Irzekevicius et al.

$$E_o = 0.394 + 0.77(1-x) + 0.2(1-x)^2$$

NOTES: electrical meas. on single crystals at 300°K

REFERENCES: Willardson et al. (A)

Temperature Coeff.

$$E_o = 1.42 - 0.98x - (4.6-1.1x)\times10^{-4}\,T$$

NOTES: optical measurement — REFERENCES: Oswald

$\frac{dE_g}{dT}$	x	
	17	$1.8\times10^{-4}\,\text{eV/°K}$

NOTES: I-V data — TEMP.: 77-300°K — REFERENCES: Ross & Snitzer

Barrier Height

x	Value	E_g (eV)	NOTES	TEMP.(°K)	REFERENCES
33	1.18 eV	1.04	photoemission meas. Cs-coated, Zn-doped, $n_n = 10^{19}$	300	Bell et al.
15	1.4 eV / 1.1	1.18	Rb-coated / Cs-coated	300	

Seebeck Coeff. ($-Q$)

x	340°C	500°C	800°C	UNIT	NOTES	REFERENCES
80	325	285	170	$\mu V/°C$	n-type, polycrystalline, $5\times10^{16}-5\times10^{17}$	Bowers et al., Weiss
90	300	260	150			
95	260	210	140			

Magnetic Susceptibility (χ_g)

x	χ_g	UNIT	TEMP.(°K)	REFERENCES
80	-0.282	10^{-6} cgs	293	Busch & Kern

Refractive Index (n)

x	n (5μ)	n (25μ)	REFERENCES
40	3.2	2.75	Oswald
60	3.3	2.9	
80	3.38	3.2	

Photocathode Properties

x	E_g	Efficiency (%)	Wavelength (μ)	Sensitivity	NOTES	TEMP.(°K)	REFERENCES
13-14	1.19	2	1.06	600μ A/lm	Cs_2O-covered on liquid epitaxy layers, 2-4μ thick	77	James et al.
15		0.8	1.06		Cs_2O-covered (Cs monolayer and Cs_2O monolayer)	77	Sonnenberg (A, B)
25		0.2	1.06				
40		0.01	1.3				
	1.04	0.1	1.2			300	Bell et al.

INDIUM-ARSENIC-PHOSPHORUS SYSTEM

PROPERTY	SYMBOL	VALUE	UNIT	NOTES	TEMP.($^\circ$K)	REFERENCES
Laser Properties	x	Wavelength (μ)	Threshold Current Density			
	6	0.942	2.5-6x10^3 A/cm^2		77	Eliseev et al.
	51	1.602	6.4-12.6x10^3 A/cm^2	halogen, vapor-deposited, 1-4x10^{19}	77	Alexander et al.
	6	0.943	5-45 for areas 1.5 to 2.5 x10^{-3} cm^2		77	Basov et al.
	20	1.09	49-66 for p-n junction areas 1-2x10^{-3}cm^2			

x	Dopant	n(10^{19})	Threshold Current Density (10^3 A/cm^2)	Wavelength (μ)	Efficiency (%)	Peak Output (W)	p-n Junction Depth(μ)	Temp. ($^\circ$K)	
7	S	1.6	2.1	0.9784-0.9837	19	1.5	15-37	80	Willardson
14	Se	0.67	4.7	1.0537-1.0601	1.3	0.66	25-27	102	et al. (A)
14.5	Te	1.5	3.7	1.0190-1.0298	21	6.8	38	102	
16.5	Te	0.19	6.7	1.0567-1.0597	0.83	0.60	28	102	
17.5	Se	0.25	8.4	1.0750	0.65	0.57	53	102	
17.0	Zn	1.0	3.07	1.0638	20	0.1	8	102	Ross & Snitzer

Emission Coeff. $d\lambda/dT$ = 1.6 Å/$^\circ$K

ALEXANDER, F.B. et al. Spontaneous and Stimulated Infra-Red Emission from Indium Phosphide Arsenide Diodes. APPLIED PHYS. LETTERS, v. 4, no. 1, Jan. 1, 1964. p. 13-15.

ALLEN, H.A. and E.W. MEHAL. Deposition of Epitaxial $InAs_xP_{1-x}$ on GaAs and GaP Substrates. ELECTROCHEM. SOC., J., v. 117, no. 8, Aug. 1970. p. 1081-1082.

ANTYPAS, G.A. and T.O. YEP. Growth and Characterization of Liquid-Phase Epitaxial $InAs_{1-x}P_x$. J. OF APPLIED PHYS., v. 42, no. 8, July 1971. p. 3201-3204.

BASOV, N.G. et al. Properties of Injection Lasers at 0.8-1.1 μ. SOVIET PHYS. TECH. PHYS., v. 12, no. 2, Aug. 1967. p. 250-257.

BELL, R.L. et al. Interfacial Barrier Effects in III-V Photoemitters. APPLIED PHYS. LETTERS, v. 19, no. 12, Dec. 15, 1971. p. 513-515.

BOWERS, R. et al. $InAs_{1-x}P_x$ as a Thermoelectric Material. J. OF APPLIED PHYS., v. 30, no. 7, July 1959. p. 1050-1054.

BRAUNERSREUTHER, E. et al. Hall Constant and Electron Mobility of InSb, InAs, and In $(As_{0.8}P_{0.2})$ in Magnetic Fields (In Ger.). ZEIT. FUER NATURFORSCH., v. 15a, no. 9, Sept. 1960. p. 795-799.

BUSCH, G.A. and R. KERN. The Magnetic Properties of the III-V Compounds (In Ger.). HELVETICA PHYSICA ACTA, v. 32, no. 1, Mar. 10, 1959. p. 24-57.

CARDONA, M. et al. Electroreflectance at a Semiconductor-Electrolyte Interface. PHYS. REV., v. 154, no. 3, Feb. 1967. p. 696-720.

DUBROVSKII, G.B. Dependence of the Forbidden Energy Gap Width of the Compound InP_xAs_{1-x} on Composition. SOVIET PHYS. SOLID STATE, v. 5, no. 3, Sept. 1963. p. 699-700.

EGOROV, L.A. and O.D. TORBOVA. Crystallization of the Indium Arsenic Phosphide Solid Solution from the Gas Phase (In Russ.). AKAD. NAUK SSSR. IZV. NEORGAN. MAT., v. 5, no. 1, 1969. p. 173-174.

EHRENREICH, H. Electron Mobility of Indium Arsenic Phosphide. J. OF PHYS. AND CHEM. OF SOLIDS, v. 12, no. 1, 1959. p. 97-104.

ELISEEV, P.G. et al. Coherent Radiation from p-n Junctions in Indium Arsenide-Phosphide. SOVIET PHYS. SOLID STATE, v. 8, no. 4, Oct. 1966. p. 1025-1026.

FOLBERTH, O.G. Mixed Crystal Formation in $A^{III}B^V$ Compounds (In Ger.). ZEIT. FUER NATURFORSCH., v. 10a, 1955. p. 502-503.

GIESECKE, G. and H. PFISTER. Precision Determination of the Lattice Constants of III-V Compounds. ACTA CRYSTALLOGRAPHICA, v. 11, 1958. p. 369-371.

IRZIKEVICIUS, A. et al. Electroreflectance Studies of $\Gamma_{15}\vec{\Gamma}_1$ Transitions in $InAs_{1-x}P_x$ Alloys. PHYS. STATUS SOLIDI B, v. 49, 1972. p. K87-K89.

JAMES, L.W. et al. Optimization of the $InAs_xP_{1-x}$-Cs_2O Photocathodes. J. OF APPLIED PHYS., v. 42, no. 2, Feb. 1971. p. 580-586.

KAVALIAUSKAS, J. and A. SILEIKA. Electroreflectance Studies of $\Lambda_3 \to \Lambda_1$ Transitions in $InAs_{1-x}P_x$ Alloys. PHYS. STATUS SOLIDI, v. 38, no. 1, Mar. 1970. p. K73-K76.

KROITORU, S.G. et al. Energy Structure of Several Solid State Materials on a Basis of Combinations of Groups III-V (In Russ.). AKAD. NAUK SSSR. IZV. NEORGAN. MAT., v. 2, no. 5, 1966. p. 805-809.

MAKHALOV, Yu.A. and R.L. MELIK-DAVTYAN. Measurement of Effective Mass of Conduction Electrons in Solid Solutions of the InAs-InP System by the Faraday Effect. SOVIET PHYS. SOLID STATE, v. 11, no. 9, Mar. 1970. p. 2155-2157.

OSWALD, F. Optical Investigations of the Semiconducting Mixed-Crystal Series $In(As_yP_{1-y})$ (In Ger.). ZEITSCHRIFT FUER NATURFORSCH., v. 14a, no. 4, Apr. 1959. p. 374-379.

PIDGEON, C.R. et al. Electroreflectance Study of Interband Magneto-Optical Transitions in Indium Arsenide and Indium Antimonide at 1.5°K. SOLID STATE COMM., v. 5, no. 8, Aug. 1967. p. 677-680. [A]

PIDGEON, C.R. et al. Interband Magnetoabsorption in Indium Arsenide and Indium Antimonide. PHYS. REV., v. 154, no. 3, Feb. 1967. p. 737-742. [B]

ROSS, B. and E. SNITZER. Optical Amplification of 1.06-μ $InAs_{1-x}P_x$ Injection-Laser Emission. IEEE J. OF QUANTUM ELECTRONICS, v. QE-6, no. 6, June 1970. p. 361-366.

SHAKLEE, K.L. et al. Electroreflectance and Spin-Orbit Splitting in III-V Semiconductors. PHYS. REV. LETTERS, v. 16, no. 3, Jan. 1966. p. 48-50.

SONNENBERG, H. InAsP-Cs$_2$O, A High-Efficiency Infrared-Photocathode. APPLIED PHYS. LETTERS, v. 16, no. 6, Mar. 15, 1970. p. 245-246. [A]

SONNENBERG, H. Long-Wavelength Photoemission from InAs$_{1-x}$P$_x$. APPLIED PHYS. LETTERS, v. 19, no. 10, Nov. 15, 1971. p. 431-433. [B]

THOMPSON, A.G. et al. Preparation and Optical Properties of InAs$_{1-x}$P$_x$ Alloys. J. OF APPLIED PHYS., v. 40, no. 8, July 1969. p. 3280-3288.

TIETJEN, J.J. et al. The Preparation and Properties of Vapor-Deposited Epitaxial InAs$_{1-x}$P$_x$ Using Arsine and Phosphine. ELECTROCHEM. SOC., J., v. 116, no. 4, Apr. 1969. p. 492-494.

UGAII, Ya.A. et al. Equilibrium Vapour Pressure Over Fused Solid Solutions of Indium Phosphide and Arsenide (In Russ.). AKAD. NAUK SSSR. IZV. NEORGAN. MAT., v. 3, no. 9, 1967. p. 1555-1560.

VAN DEN BOOMGAARD, J. and K. SCHOL. The P-T-x Phase Diagram of the Indium-Arsenic, Gallium-Arsenic and Indium-Phosphorus Systems. PHILIPS RES. REPTS., v. 12, no. 1, Apr. 1957. p. 127-140.

VISHNUBHATLA, S.S. et al. Electroreflectance Measurements in Mixed III-V Alloys. CANADIAN J. OF PHYS., v. 47, no. 16, Aug. 1969. p. 1661-1670.

WEISS, H. Thermoelectric Power and Heat Conduction of III-V Compounds and Their Mixed Crystals (In Ger.). ANNALEN DER PHYSIK, v. 4, no. 1-5, 1959. p. 121-131.

BELL AND HOWELL RES. LABS., PASADENA, CALIF. Indium Arsenide-Phosphide Injection Lasers. By: WILLARDSON, R.K. et al. Contract N00014-68-C-0219. Nov. 1968. 118 p. AD 679 908. [A]

BELL AND HOWELL RES. LABS., PASADENA, CALIF. Semiconductor Materials for Electroluminescence Diodes and Lasers. Tech. Summary Rept. By: WILLARDSON, R.K. et al. Contract no. N123 (62738)55582A. May 1968. AD 709 970. [B]